ROUTLEDGE LIBRARY EDITIONS:
MEDIEVAL CULTURE,
SOCIETY AND RELIGION

Volume 11

GEOGRAPHY IN THE MIDDLE AGES

GEOGRAPHY IN THE MIDDLE AGES

GEORGE H. T. KIMBLE

Routledge
Taylor & Francis Group
LONDON AND NEW YORK

First published in 1938 by Methuen & Co. Ltd.

This edition first published in 2025
by Routledge
4 Park Square, Milton Park, Abingdon, Oxon OX14 4RN

and by Routledge
605 Third Avenue, New York, NY 10158

Routledge is an imprint of the Taylor & Francis Group, an informa business

British Library Cataloguing in Publication Data
A catalogue record for this book is available from the British Library

ISBN: 978-1-041-02220-6 (Set)
ISBN: 978-1-041-00716-6 (Volume 11) (hbk)
ISBN: 978-1-041-00718-0 (Volume 11) (pbk)
ISBN: 978-1-003-61128-8 (Volume 11) (ebk)

DOI: 10.4324/9781003611288

I

MACROBIAN VIEW OF THE WORLD (c. 1485)
(from '*Macrobii . . . in somnium Scipionis expositio*')

GEOGRAPHY
IN THE MIDDLE AGES

by

GEORGE H. T. KIMBLE

Lecturer in Geography in
The University of Reading

With 20 Plates

LONDON
METHUEN & CO. LTD.
36 Essex Street W.C.2

First published in 1938

PRINTED IN GREAT BRITAIN

PREFACE

WHILE there have been many histories of geographical exploration, the history of geographical thought has still to be written. This volume is offered as a modest contribution to that study. It essays to follow the development of medieval earth-knowledge as it was affected, on the one hand, by the prevailing ideologies and, on the other, by the results of human enterprise. Much of the ground covered will be familiar to students, both of history and geography, for my aim has been to select representative men and writings to illustrate the theme rather than to catalogue all the available material, literary and cartographical. Further, few of the conclusions will strike the well-informed reader as novel, but I comfort myself with the knowledge, firstly, that they have not previously been made accessible within the compass of a single volume, and secondly, that the contribution of the Middle Ages to the advancement of geographical studies has still to be adequately appreciated.

In preparing this work I have enjoyed the help of many friends whose scholarship I have sought to put to good account. In particular, I should like to name Professor E. G. R. Taylor and Mr. W. F. Morris, who read the manuscript and made many constructive criticisms ; Professor T. E. Jessop and Mr. J. N. L. Baker, who revised the proofs and rescued me from several of the pitfalls which beset the unwary author ; Messrs. L. R. Harris and J. R. Casswell, who gave invaluable assistance in the translation of difficult Latin passages ; and, not least, to Mr. Robert Gibbings, Mr. Peter

Wait and the members of the Reading University Library staff for numerous incidental courtesies.

For permission to reproduce the plates contained in the book, I am indebted to the Trustees of the British Museum, the Council of the Hakluyt Society, and Dr. Richard Uhden of Berlin.

Lastly, my thanks are due to the University of Reading for a generous grant allocated to me for this work.

READING UNIVERSITY
January 1938

CONTENTS

ILLUSTRATIONS

GEOGRAPHY IN THE MIDDLE AGES

CHAPTER I

THE PASSING OF CLASSICAL GEOGRAPHY

GEOGRAPHY is notoriously the Cinderella of the sciences. Although one of the oldest, it has been treated, until recent years, with unbecoming and wholly unmerited disrespect. Modern universities are not alone in their reluctance to accord the subject full academic honours. The Ancients, for all their enlightenment, could not make up their minds about its significance. One school—the Platonists—held that it should be considered as a branch of physics. Another—the Aristotelians—averred that, as it was primarily concerned with measurement, it ought to be regarded as a branch of applied mathematics, like geometry or astronomy. The scholars of the Middle Ages, the intellectual inheritors of the Ancients, were equally embarrassed. Geography, they affirmed, could not reasonably be fitted into the Quadrivium [1] or the Trivium.[2] So it came to stand for the odds and ends of knowledge unaccommodated by the seven liberal arts. Nevertheless, the subject did not go unchampioned, for the medieval theologian, with his vested interest in the origin, shape, movement and peopling of the earth, found it a most fruitful field for the dissemination of the Church's doctrines. Nor did the theologian's interest in the subject stop here. To him every created thing was endowed with allegorical significance. Just as man's body exists for his spirit's sake, so—the argument

[1] The four mathematical disciplines—arithmetic, music, geometry and astronomy. The word seems to have been first popularized by Boethius, the Roman philosopher and statesman of the early sixth century A.D.

[2] The remaining three of the seven recognized liberal arts, viz. grammar, dialectic and rhetoric.

went—in creating the world for man, God had in view his most important interests, which were not those of his mortal body, but those of his immortal soul. Consequently the creation and ordering of the world was held to subserve man's spiritual interests and the divine account of it to carry spiritual instruction. 'The spirit was created for God's sake; the body for the spirit's sake, and the world for the body's sake, so that the spirit might be subject to God, the body to the spirit, and the world to the body.'[1] From which it followed that all human knowledge, in handmaidenly fashion, could be held to promote an understanding of man as well as of the teaching contained in Scripture. Accordingly, almost every encyclopaedia coming from the medieval monasteries—and most of the scholarship of the age was clerical—has a geographical, or rather a cosmographical section.[2] But cosmography meant different things to different men. To some it was primarily the description of the creation of the world and the spread of mankind across its face; to others it was primarily the study of the phenomena of weather, climate, plants, animals, precious stones and 'wonders'; while to others it was more nearly what we understand by the term 'political history'. However, within this wide field there was much common ground; for this reason it is possible to grasp, without much difficulty, the scope of world-knowledge in a particular period of the Middle Ages.

The time of the beginning of the Middle Ages is obscure. Obviously it is impossible to assign a date; in some respects the Fathers of the Church may be said to have opened the era, yet the pagan classical civilization continued for many centuries, Even the closing of the School of Athens in A.D. 529 is not a sufficient landmark, although by that time the Christian academies of Alexandria, Beirut and Constantinople were certainly the more influential. It is more satisfactory to think of this age—as indeed of any other conventional time-period—

[1] Hugo of St. Victor: Prologue to *De Sacramentis Christianae Fidei.*

[2] The word 'geography' is absent from ordinary usage throughout the thousand years of the Middle Ages.

as overlapping the previous age rather than as discrete from it. For our immediate purpose, a convenient dividing line may be drawn in the reign of Constantine (*c.* A.D. 300). From his time onward until the Columbian era, European civilization was stamped with the imprint of two traditions and disciplines—the classical and the Christian. These were the hall-marks of the period, for however we try to define it, we cannot lose sight of the fact that 'medieval Europe delighted to think of itself as the old Roman world state under religion.'[1]

But this is not to say that in point of fact the men of the Middle Ages shared the intellectual and geographical background of the Ancients. Nothing could be further from the truth. In order to see exactly how far removed from Antiquity was the typical medieval conception of the *oikoumenē* or habitable world, to realize how limited were the horizons even of the most cultured men by comparison with those of the previous age, and to appreciate the *naïveté* of mind which did not question the corporeal existence of dragons and of men without heads, we should remind ourselves, first of all, of the world-outlook at the close of the classical era.

What was the 'world' of Constantine's day? The only contemporary eye-witness descriptions are those found in the itineraries—*itineraria annotata*—which are few in number. The chief of these are the *Itinerarium provinciarium Antonini Augusti*, written early in the third century, but edited and enlarged later, the *Bordeaux-Jerusalem Itinerary*, written A.D. 333 by an unknown Christian pilgrim,[2] and the *Pilgrimage of Etheria*, the record of a journey made in A.D. 385 by Etheria (or Silvia) of Aquitania to Jerusalem.[3] All three documents are disappointing. The writers of the latter two were otherworldly; the visiting of Biblical sites was their first aim and they pass over, with only the briefest mention, their routes to and from the Holy Land.

[1] C. R. Beazley: *The Dawn of Modern Geography*, Vol. I, p. 9.

[2] *Vide* English Translation by A. Stewart (Palestine Pilgrims' Text Society, Vol. I, 1887.).

[3] *Vide* English Translation by M. L. McClure and C. L. Feltoe, 1920.

The *Bordeaux Itinerary*—the first pilgrim record extant—is typical of their method and scope. It commences as follows:

An Itinerary from Bordeaux to Jerusalem, and from Heraclea (Eregli) through Aulon (Avlona) and through the city of Rome to Milan. As follows:

The city of Bordigala (Bordeaux) where is the river Garonne, in which the ocean ebbs and flows for a hundred leagues (Gallic league = 1½ Roman miles), more or less.

Change at Stomatae (Castres)	VII	leagues
Change at Senone (Sirio, Pont de Ciron)	IX	,,
city of Vasates (Cossio, Bazas)	VIII	,,
Change at Three Trees	V	,,
Change at Oscineium (Honeilles)	VIII	,,
Change at Scotium (Sotium, Sos)	VIII	,,
Change at Vanesia	XII	,,
city of Auscuis (Augusta, Auscorum, Auch)	VIII	,,
Change at the sixth league	VI	,,
Change at Hungunverrum	VII	,,
Change at Buccones (Isle en Jourdain)	VII	,,
Change at Temple of Jupiter	VII	,,
city of Tolosa (Toulouse)	VII	,,

Once within the confines of the Holy Land the itinerary is amplified to give thumbnail sketches of all the places of historical—mainly Biblical—importance.

Such statements as these are of interest to us chiefly as commentaries on the Imperial road-system of the time. The *Antonine Itinerary* is important for much the same reason. It describes the routes and administrative divisions of the Empire after their reorganization by Diocletian and so furnishes us with materials for geography, but not with a geographical system. Moreover, its scope is confined strictly to the limits of the Roman Empire. In Mauretania, for instance, we find no town on the western coast beyond Sala, although there was an outpost, Mercurius by name, 16 miles farther south. Towards the northern frontier line, Hadrian's wall provides the limit, save for an outlying station called Bremenium, 20 miles to the north of it. In Gaul we find the Rhine forming the absolute limit on the east, as did the Danube towards the north throughout its course, from Guntia near Ulm to Noviodunum within a few miles of its mouth. On the eastern side, the itinerary embraces routes as far distant as Samosata on the Euphrates and on the southern front the *ne plus ultra* is supplied by the outpost of Hiera Sycaminos, near Dakkah, in the valley of the Nile.

The *Tabula Peutingeriana* (so called from the name of its former owner, the archaeologist Konrad Peutinger of Augsburg) is an itinerary of a different kind. Drawn probably about the middle of the third century, it is an attempt to represent the Empire cartographically as it was in the first century, with the exception of a few Christian additions.[1]

It is obvious that these works, while helping to frame the picture of the fourth-century world, do little towards filling it. Unfortunately, the 'geographers' of the time are hardly more helpful. The third and fourth centuries only bequeath the names of three—Julius Solinus, surnamed *Polyhistor*,[2] Avienus[3] and Pappus[4]—and none of these described the world of his own day. Solinus' *Collectanea Rerum Memorabilium* is a work of very little real geographical value (yet one, nevertheless, of almost unrivalled popularity in the Middle Ages). What Solinus did was to put together, from all available sources, a number of marvellous tales—tales of strange animals, monstrous races, curious stones and of natural wonders of the earth and sea. The only thing geographical about the compilation is the method. Thus he begins by describing Italy; thence he takes us through Pontica, Scythia, the Black Sea lands and Russia. From there he goes to the northern and western countries of Europe and concludes his 'varied narration' (whence the name *Polyhistor*) by speaking of Africa and Asia. Nine-tenths of his material are taken from Pliny, whose name is not so much as mentioned; the rest comes

[1] C. R. Beazley, *op. cit.*, Vol. I, p. 383.
[2] *Fl. c.* A.D. 250. [3] *Fl. c.* A.D. 350–400. [4] *Fl. c.* A.D. 300.

from Mela and other late classical writers. In the knowledge of the earth he is far behind Herodotus, though claiming to cover a wider field. Among the very few instances where Solinus added anything to his authorities is his mention of the British Isles, concerning which there must clearly have been a greater body of common knowledge in his day than in the first century. Yet what Solinus tells us is at once scanty and elementary. The only commercial products he appears to have known of the country which was then providing the Empire with grain, wool, precious metals and cloth, were stones called 'gagates'—evidently jet—and pearls. The same slightness characterizes his descriptions of the eastern fringes of the habitable earth. About 'India extra Gangem' he says nothing, while his account of the land of Seres is simply a repetition of Pliny. 'The barbarous style of our author with his perpetual distortions and attempts at rhetorical grandiloquence . . . doubtless contributed greatly to the favour which his work enjoyed during the centuries which followed its publication'.[1]

Concerning Pappus very little is known beyond the fact that he was a Greek geometrician who lived in Alexandria and wrote a text-book of geography—*Chorographia Oikoumenikē*—which has been lost. However, its substance may be gathered from the use made of it by Moses of Chorene (*fl. c.* 440) who says that Pappus had based his work on Ptolemy's maps. This contention is fully borne out when we compare the text of Moses' work with the earliest Ptolemaic maps we know. Only so, in fact, can the origin of that text be explained, for while its general arrangement conforms well with Ptolemy's maps, it shows little resemblance to Ptolemy's text.

Avienus tried to combine the functions of geographer and poet. One of his works—*Descriptio Orbis Terrae*—is a metrical paraphrase of a popular second-century work, entitled *Periegesis*, by Dionysius. Only very occasionally does the poet depart from his original and then scarcely ever to

[1] E. H. Bunbury: *History of Ancient Geography*, Vol. II, p. 678.

advantage. In describing the Alps and their river system he does introduce some original material, but this was not difficult, for Dionysius knew little beyond the name. The lines portraying them are characteristic of his general method :

> Nec procul hinc rigidis insurgunt rupibus Alpes,
> Nascentem diem celso juga vertice cernunt.
> Porro inter cautes et saxa sonantia Rhenus.
> Vertice qua nubes nebulosus fulcit Adulas,
> Urget aquas, glaucoque rapax rotat agmine molem
> Gurgitis, Oceani donec borealis in undas
> Effluat et celeri perrumpit marmora fluctu. V. 433.[1]

Avienus' more important geographical composition—*Ora Maritima*—which survives only as a fragment, is also in metrical form ; it describes the shores of the Mediterranean, Caspian and Black Seas in iambic hexameters. The lines which are extant, numbering about 700, deal only with the Ocean coasts west of the Columns of Hercules and the Mediterranean from the Straits to the mouth of the Rhone. In content they amount to little more than a list of statements of the vaguest character, strung together without connexion or method and derived from heterogeneous and mostly obsolete sources.

Where, then, can we look for our world-picture ? Are we justified in going back to the Great Age of classical geography and using Ptolemy's ? Ptolemy was still being read in the fourth century, for Ammianus Marcellinus, the Roman historian, is partially indebted to him for the geographical matter contained in his *Rerum Gestarum Libri XXXI.*[2] More-

[1] And not far hence the Alps rise up with rugged cliffs,
Their towering peaks beholding the dawn [i.e. the east].
Beyond, amid the crags and the echoing rocks,
From the summit where Adulas, veiled in mists,
Rests his clouds, the Rhine pours forth its water,
And, sparkling, swirls its mighty torrent along,
Until it flows into the waters of the Northern Ocean
And cleaves them with its strong current.
(Translation by L. R. Harris).

[2] *Vide* E. H. Bunbury : *op. cit.* Vol. II, p. 679 *et seq.*

over, Ambrosius Theodosius Macrobius [1] and Martianus Capella [2] both lean towards the Ptolemaic view of the universe.

Let us see to what extent their views tally. Macrobius' cosmography lies embedded in a commentary entitled *In Somnium Scipionis*. With its postulate of a stationary round earth at the centre of the universe and its contention that the environmental sea, variously called the Atlantic, the Great Sea and the Ocean, is 'in spite of these big names, quite small', it is definitely in the Ptolemaic tradition. It departs, however, from that tradition in making the ocean the boundary, in every direction, of the inhabited earth, giving it the shape of a lozenge, narrow at the extremes and wide in the middle,[3] and in positing the existence of three other landmasses corresponding to the *oikoumenē*, in the remaining quarters of the earth.[4] In his territorial division, Macrobius adopts the conventional five zones, and, while maintaining the existence of an Antipodean race of men, he yet holds that there is no way by which knowledge of them can be obtained, owing to the impassability of the Torrid Zone [5]—a theory which can in no way be reconciled with Ptolemy's catalogue of places in the Southern Hemisphere. Again, Macrobius differs from Ptolemy on the question of the size of the earth. Whereas Ptolemy calculated the circumference to be 180,000 stadia, i.e. 22,500 miles (reckoning 500 stadia of 625 feet to one degree) Macrobius preferred Eratosthenes' figure of 252,000 stadia, or

[1] *Fl. c.* A.D. 395–423.

[2] *Fl. c.* 470, *vide* G. Sarton: *Introduction to the History of Science*, Vol. I, p. 385.

[3] *Commentarius in Somnium Scipionis:* II, 9. *Vide* Plate No. I.

[4] *Ibid.*

[5] *Ibid.:* II, 5. Such impassable barriers between the habitable parts of the earth are used by him, as by Cicero, to show the limitations of fame in space. Similarly, he carries forward the argument from the necessity to which the earth, in his estimation, is subject, of alternate floods and fires, to prove the limitations of fame in time. The moral—for practically all knowledge, at this period, was regarded as having a moral significance—is that virtue must not be practised in order to acquire fame.

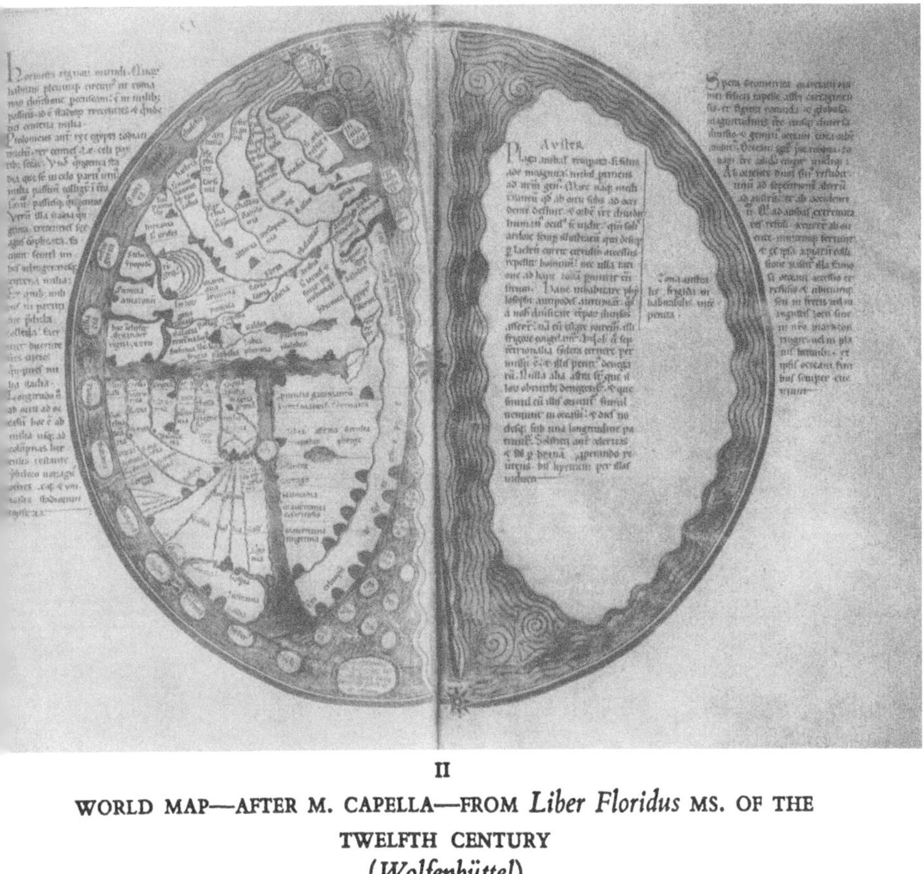

II

WORLD MAP—AFTER M. CAPELLA—FROM *Liber Floridus* MS. OF THE TWELFTH CENTURY

(*Wolfenbüttel*)

25,000 miles approximately (based upon a reckoning of 700 stadia of 500 feet to one degree).[1]

Capella's views are found in the Sixth Book—on Geometry —of his *Satyricon.*[2] First, he deals with the earth's form. This he regards as globular, accepting the findings of Dicaearchus, Ptolemy and Eratosthenes. In his estimate of the earth's circumference he shares Macrobius' preference for Eratosthenes' figure. He then proceeds to the division of the earth into five zones; three of these are 'intemperate by an excess of contrary qualities',[3] two being uninhabitable from cold and one from heat. On the extent of the northern temperate zone he refers to the authority of Ptolemy and Pytheas [4]—'most learned of men'—but declares that his own knowledge is no whit inferior to theirs, 'for,' says Capella, 'I, too, have been all over these [countries], nor does any part of the earth, as I think, remain unknown to me'.[5] Contrary to Ptolemy, he holds the Ocean to be entirely circumnavigable.[6] The voyage of Augustus Caesar from Germany to Scythia attested this, he says, so far as the Northern seas were concerned [7]; the voyages of the Macedonians from India into the Caspian Sea (*sic*) did the same for Eastern waters,[8] while for Southern and Western waters come the equally clear

[1] i.e. Ptolemy reckoned 8 ordinary *land* stadia and Eratosthenes 10 *sea* stadia to the mile. It was on the authority of Macrobius, of whose writings in Latin every man of learning had knowledge (in the Middle Ages), that the value of Eratosthenes was made so widely current. Sacrobosco was one of the many later-day scholars to utilize it, and his work *De Sphaera* became one of the chief nautical text-books of the early Portuguese shipmen.

[2] Or *De Nuptiis Philologiae et Mercurii et de septem artibus liberalibus libri novem* (Edited by F. Eyssenhardt, 1866). In a recent monograph, entitled *Die Weltkarte des Martianus Capella*, R. Uhden has pointed out that the world map contained in the twelfth century manuscript of Lambert of St. Omer's *Liber Floridus* (preserved at Wolfenbüttel) carries a legend ascribing the original to Capella. The correctness of the ascription is fully vouched for by the identity of various legends on the map with passages in the *De Nuptiis* . . . *Vide* Plate No II.

[3] *Ibid.*: VI, 602. [4] Some editions give Pythagoras. [5] *Ibid.*: VI, 609.

[6] *Ibid.*: VI, 619. [7] *Satyricon* . . .: VI, 617–18. [8] *Ibid.*: VI, 619.

testimony of the voyages of Hanno, Eudoxus and the unnamed trader mentioned by C. Antipater.[1]

Now it is clear from such statements that, by the fourth century, geographical thought was no longer being moulded on the Ptolemaic pattern. Further investigation shows that the opinions of Herodotus, Eratosthenes, Hipparchus and Strabo were no longer accepted, as they had been hitherto, as authoritative. In their stead, the *De Situ Orbis* of Pomponius Mela and the *Natural History* of C. Pliny, palpably poorer works in a modern estimation, were promoted to positions of high regard. In fact after the fifth century and until the beginning of the fifteenth, we find only occasional references to Ptolemy and his school, while, as for his *Geography*, it is scarcely ever noticed.[2]

Does this mean, therefore, that Ptolemy's 'world' had contracted by Constantine's time? The work of a man like Capella leaves us in little doubt as to the answer. As we have seen, he follows Pliny and Solinus, rather than Ptolemy, and, significantly, those chapters which are relevant to our purpose are at once meagre in factual detail and furnished instead with that geographical mythology which throughout the Middle Ages offered so sturdy a resistance to scientific progress: the perpetual motion of the Hyperboreans who inhabit the lands of the North 'where the axis of the world is ever turning'[3]; the reddening fountain on the shores of the Red Sea[4]; the marvels of India where men reach fabulous height and age, worship Hercules and never sleep by day[5]; the iron beams barring the entrance of the Caspian Gates[6]; Mount Atlas 'rising to the confines of the lunar circle beyond the power of the clouds,[7] and Thule, near which the sea is always

[1] *Satyricon . . . .:* VI, 620–2.

[2] *Vide infra*, chap. ix.

[3] *Satyricon . . .:* VI, 664, i.e. the axis on which the celestial, as well as the terrestrial, sphere was supposed to turn.

[4] *Ibid.:* VI, 677. Reddened by desert sand?

[5] *Ibid.:* VI, 697. [6] *Ibid.:* VI, 691.

[7] *Ibid.:* VI, 667. The fact that high mountains project up through the cloud-belt had been observed by Aristotle: *Meteorologica*, I, 3 *et seq.*

frozen.[1] Here also we find the habitable globe confined again to the Northern Hemisphere. The southern extension of Africa, established by Ptolemy (and perhaps Herodotus), is ignored, and the Nile is again made to cross that Continent from west to east in a line parallel with the Southern Ocean:[2] Egypt is regarded as forming part of Asia [3] and Africa commences only to the west of the Nile. The Ganges flows into the eastern ocean and the Caspian (the true character of which Ptolemy, following Herodotus, had reasserted after centuries of misconception) is once more held to be a gulf of the ocean. Errors and confusions such as these abound in the *Satyricon* and, for the most part, are perpetuated by the other writers of the age.

Here, then, is evidence in plenty of both a contracting horizon and a decadent geography. So far from eliciting new items of geographical information, the works of this age suffer from errors and limitations characteristic of the pre-Ptolemaic writings. Scholarship is more and more restricted in its range and ceases to be cultivated for its own sake. From the fourth century onwards many writers considered geography to be of importance only in so far as it bears on Biblical sites and place-names, and map-making only in so far as it gives them a local habitation. The popularity enjoyed by Capella and Macrobius in the following centuries—in the twelfth century the *Commentary* was second only in favour to the *Satyricon*, often the only text-book used in schools [4]—was a sufficient barrier to the resurrection of scientific geography,

[1] *Ibid.:* VI, 666. If Thule is Iceland, then this statement contains a germ of truth.

[2] *Ibid.:* VI, 676. This idea goes back to Herodotus, *vide History*, Book II, chap. 31.

[3] *Ibid.:* VI, 695. Cf. Pomponius Mela : *De Situ Orbis*, Book I, chap. 8, and C. Pliny : *Natural History*, Book III, chap. 3.

[4] Capella was regarded with disfavour by those Christian teachers who rigorously proscribed pagan literature ; and his book, though it remained an authority down to the Renaissance, was not everywhere admitted. Thus it is absent from Alcuin's catalogue (*c.* A.D. 770) of the library at York, a fairly representative collection of the books which were then most read. *Vide* Sir Richard Jebb in *Cambridge Modern History*, Vol. I, p. 533.

though it must be admitted that these two works did serve to keep alive belief in the sphericity of the earth.

The causes of this calamitous contraction of earth-knowledge are not far to seek. In the first place there was the fact of Rome's commercial decline. Under the early Roman Empire, the trade with India, the Persian Gulf, Arabia and the eastern coast of Africa had been in the hands of Roman merchants who sailed the Red Sea and Indian Ocean in their own vessels. By the end of the third century, the forces of disruption had so far gripped the Empire that it was no longer the sole guardian of these great trade routes. Traffic with the Far East slowly passed into the hands of intermediaries, Persians, Abyssinians and Himyarites of Yemen. The Persians secured a monoply of the Chinese silk trade while the Abyssinians and the Himyarites became the main carriers of the Indian trade. African trade, mainly in slaves, ivory and precious metals, also fell into the hands of the Abyssinians. The Roman traffic with trans-Alpine Europe suffered a like gradual eclipse. The importance of this for the Imperial economy was far-reaching, but it is hardly our concern. For geography, the cessation of direct intercourse with the pagan world, particularly the Far East, meant a very far-reaching restriction of its territorial frontiers, and, consequently, a loss of knowledge. This is illustrated in such ways as the misuse of the name India to designate Ethiopia, a common practice of Greek and Latin writers from the fourth century.[1] To look for expanding horizons and fuller knowledge in such circumstances is merely to court disappointment. Where there is no vision, knowledge and enterprise perish.

In the second place there were the Germanic invasions of the fifth and sixth centuries, which undermined the unity of Roman life and culture. They caused not only the reduction of the exchange of products and ideas outside the Empire to a minimum, but also, because of the consequent collapse of the Imperial administrative system, the breakdown of political and commercial life within the Empire. Ideas and information,

[1] *Vide* J. B. Bury : *History of the Later Roman Empire*, Vol II, p. 318.

henceforth, spread only slowly and against great resistance from one district to another; culture became regional and stagnant.

The reorganization of the Roman Empire into two parts by Theodosius at the end of the fourth century, to cope with the new menace, while sound strategically, served only to further these tendencies; for, under the new order, the Latin West and the Greek East gradually drifted asunder. After the schism of A.D. 476, when the two empires became separate entities, the contacts of the Latin world with Greek civilization became more and more tenuous, to the incalculable loss of the West. From the fifth to the twelfth century Greek scholarship might not have existed—so little influence did it exert on Western culture.

But external factors, such as the invasions, important though they were in this respect, should not blind us to the operation of internal forces. Indeed, it is doubtful whether any factor was more potent in the decline of ancient science and in the silencing of the scientific spirit than the utilitarian bias of the Latin mind. Even in the hey-day of their prosperity the Romans manifested remarkably little academic interest in the lands they had conquered. Their geographers can be numbered on the fingers of both hands, and the greatest of them, Strabo, was a native of Amasia, a town in a remote province of Asia Minor outside which he was unknown.[1] Moreover, he wrote in Greek, which few Romans could understand. As for Pliny, he was more interested—the title of his work notwithstanding—in the 'unnatural' aspects of natural history than in plain geographical descriptions. The plodding ways of scientific investigators made little appeal to

[1] His *Geography* remained unknown to most of his successors. His name is not even noticed among the multifarious writers cited by the arch-compiler Pliny: nor is any allusion to it found in the great work of Ptolemy. It was not until the time of Marcianus of Heraclea—a writer who cannot be placed earlier than the third century—that his work was first acclaimed. But his references are entirely isolated and it is only during the later years of the fifth century—the time of Stephanus of Byzantium—that he begins to receive considerable posthumous recognition.

men who thought in terms of world-dominion and of a commerce nearly as extensive.

To make matters worse, Roman utilitarianism was followed by an ecclesiastical expediency which bade fair to destroy all hope of a genuine scientific revival. During the early centuries of the Christian Church, by many of its principal authorities, scientific and philosophical inquiry was not only disregarded, but discouraged. The great theological doctrines which were presented to men's minds, and the weighty matters of the regulation of the will and the affections, which religion laid upon them, made scientific investigation seem 'a reprehensible misapplication of human powers'.[1] Many of the Fathers went so far as to revive the opinion of Socrates that the only valuable pursuit is that which teaches us our moral duties and religious hopes. The mind of Pope Gregory I,[2] for instance, was closed to all that is irrelevant to the knowledge of God and the soul.[3] Secular study and the whole round of man's mortal interests lay outside the field of his intellectual endeavour. The letter which he wrote on the subject of classical studies to Desiderius, Bishop of Vienne, is characteristic of his attitude.

> A report has reached us which we cannot mention without a blush, that thou expoundest grammar to certain friends; whereat we are so offended and filled with scorn that our former opinion of thee is turned to mourning and sorrow. The same mouth singeth not the praises of Jove and the praises of Christ. . . .[4]

Lactantius, another of the early Fathers, declares that science is both foolish and false.

> For to investigate or wish to know the causes of natural things —whether the sun is as great as it appears to be, or is many times greater than the whole of this earth; also whether the moon be

[1] W. Whewell: *History of Inductive Sciences*, Vol. I, p. 194.

[2] b. 540—Pope 590—d. 604.

[3] Cf. St. Augustine: *Soliloquiorum libri duo*, I, 2. 'Deum et animam scire cupio. Nihilne plus? Nihil omnino.'

[4] *Epistolae*: XI, 54.

spherical or concave ; and whether the stars are fixed to the heaven, or are borne with free course through the air ; of what magnitude the heaven itself is, of what material it is composed ; whether it is at rest and immovable, or is turned round with incredible swiftness ; how great is the thickness of the earth, or on what foundations it is poised and suspended—to wish to comprehend these things, I say, by disputation and conjectures, is as though we should wish to discuss what we may suppose to be the character of a city in some very remote country, which we have never seen, and of which we have heard nothing more than the name.[1]

And Lactantius was a typical product of his Church : for there were few from his time to that of Odo of Cluny who did not profess to despise the *philosophical* work of the ancients.

'What concord hath Christ with Belial ?' asks St. Jerome. 'How can Horace go with the Psalter, Virgil with the Gospels, Cicero with the Apostle ? Is not a brother made to stumble if he sees you sitting at meat in an idol's temple . . . ? We ought not to drink the cup of Christ, and at the same time, the cup of devils.'[2]

'Why cite me the rites of our ancestors ?' inquires St. Ambrose of Symmachus who, in a memorial on the proposed removal of the pagan statue of Victory from the Forum in Rome, had made a plea for its retention on grounds of tradition. 'There is no shame in passing to better things.'[3] All the same this contempt was perhaps more feigned than real, for these men, the avowed opponents of pagan culture, were the very men to keep it alive. They wrote in Latin, freely employing quotations from classical authors ; they taught Latin in their schools, and the rules and examples were essentially those of classical practice. Then, as now, the training of students was based on anthologies and text-books,

[1] *Divinae Institutiones :* III, 3 (W. Fletcher's Translation, Vol. XXI, Ante-Nicene Christian Library.).

[2] *Epistolae :* XXII, 29. Translation by W. H. Freemantle (Vol. VI, Nicene and Post-Nicene Fathers).

[3] *Ibid. :* XVIII, 7. Translation by H. De Romestin (Vol. X, Nicene and Post-Nicene Fathers).

the material for which was culled from the standard works of Antiquity. If we add to this the poor quality of much of the patristic writing, unsatisfying to intellect and heart alike, we can readily understand the attraction for the learner of classical literature.

It will, perhaps, be asked, what had the Church to do with a pagan culture? To this question the work of Lactantius provides an unexpected answer. Through the seven books of his *Divine Institutes* he is constantly tapping classical resources, using them both critically and sympathetically. From first to last he makes it quite clear that while the Christian faith finds much in pagan belief and pagan morals to avoid, it may, or rather, must draw freely for its sustenance on the thought, the poetry and the inspiration of the past. Many a man endorsed this view during the following centuries, if the popularity of the *Institutes* is any criterion; for, beginning with the revival of learning in the time of Charlemagne, it became one of the most copied works in Christendom and was the first dated book printed in Italy. By the sixteenth century there were no less than thirty-six printed editions of it. Even in his own life-time it was found so useful that an epitome of it was prepared, possibly by the author himself.

In taking up this attitude Lactantius was, moreover, highly commended by St. Augustine, Cassiodorus and Isidore, even by St. Jerome who, notwithstanding his plea, quoted above, for literary separation, maintained, at a later period of his life (*c.* A.D. 400) that the best in ancient literature and thought ought to be adapted to the Christian faith. In writing to one Magnus, an orator of Rome,[1] he says:

> You must not adopt the mistaken opinion, that while, in dealing with the Gentiles, one may appeal to their literature, in all other discussions one ought to ignore it: for almost all the books of all these writers . . . are extremely full of erudition and philosophy.

Further, he cites Biblical authority for the spoiling of the Egyptians—provided there is no pollution from the spoils.

[1] *Epistolae:* LXX, 6.

For who is there that does not know that both in Moses and in the prophets there are passages cited from Gentile books. . . . The Apostle Paul also, in writing to Titus, has used a line of the poet Epimenides.[1] In another Epistle Paul quotes a line of Menander.[2]

Nor did St. Jerome fail to practise what he preached; for in his monastery at Bethlehem, he and his fellow-monks applied themselves with great assiduity to the task of copying and translating classical works as well as the Scriptures. This magnificent example of cloistered scholarship—the first—was followed up by succeeding generations of scholars. In the next century Cassiodorus (*fl.* A.D 500–560) founded a monastery at Viviers (Castellense) with the specific purpose of encouraging learning—not, of course, as an end to itself, but as a means to a better understanding of the Scriptures. His monks were trained as copyists and were urged to study cosmography, just enough, at any rate, to enable them to understand when Scripture speaks of countries, towns, mountains or rivers, and in their reading to refer, above all, to the Scriptures and the writings of Julius Honorius, Dionysius, Marcellinus (who had composed an account of Constantinople and Jerusalem) and even Ptolemy.[3] To this end he spent large sums on the purchase of manuscripts from the famous classical libraries. His determination to utilize the leisure of the convent for the preservation of divine and human learning[4] and for its transmission to posterity, is responsible for the survival of many choice relics of ancient genius. Indeed, it is widely acknowledged that the civilization of subsequent centuries, and in particular the institution of monastic libraries and schools, where the light of learning continued to shine in the Dark Ages, owed much to the foresight of Cassiodorus.

[1] 'The Cretans are always liars, evil beasts, slow bellies' (i, 12).

[2] 'Evil communications corrupt good manners' (1 *Corinthians* xv. 33).

[3] *Vide De Institutione Divinarum Literarum*, cap. 25.

[4] His main work is entitled *Institutiones Divinarum et Humanarum Literarum*. His work on the Trivium, *De Artibus et Disciplinis Liberalium Artium*, became a favourite text-book in the Middle Ages.

When all is said and done, the temper of medieval writers was not greatly different from that of the Ancients. If Christianity had made people 'other-worldly', so too had pagan philosophy—witness the Platonic dialogues on Immortality. If the early Church had failed to set the arts above the absolutes, so had Cicero. In fact, once it was seen that Christianity was the heir of a past which had strong cultural ties with the present and that to disregard this heritage altogether would be to do the Faith lasting disservice, the doctors of the Church set about re-cultivating the classical soil. In Italy the classics never ceased to form the conscious background of all intellectual life. North of the Alps, letters came as a handmaid in the train of Latin Christianity—a handmaid who was apt to assert her own value, and also to charm the minds of men. From the fifth century onwards ecclesiastical learning drew more and more of its sustenance from this source. Consequently it may be asserted that the principal means by which Roman culture was transmitted to the Middle Ages and gave rise to the distinctive features of their culture are to be ascribed to the powerful organization of the Catholic Church.[1]

[1] *Vide* C. Foligno in *Legacy of Rome* (edited by C. Bailey), chap. I.

CHAPTER II

THE DARK AGES OF GEOGRAPHY

THE attitude of the early Church towards profane studies is of great moment in any appraisal of medieval learning. As we have already seen, it was an attitude of tolerance rather than of interest, for any open confession of interest would have invited excommunication. But the interest, veiled and suspect, was there, and to it we owe a good deal of what is best in the Dark Ages—not least the transmission of classical knowledge. It is to the monasteries, after all, that we owe the safe custody, the translation, the copying and the editing of the monumental works of Antiquity. No doubt the early centuries down to about A.D. 1100 deserve the description 'Dark Ages', from the point of view of intellectual and scientific progress (although at least one modern scholar affirms that 'those ages were never so dark as our ignorance of them'),[1] but they would have been darker still without the dim religious light of the cloisters. And what if, for a time, they did hide their light? With whom, in the crumbling, chaotic world, could the scholars of the Church have carried on intellectual traffic? It cannot be denied that many of them made only poor and eclectic use of their cultural inheritance. Their explanations of natural phenomena, of physical forces, of the size and motions of the heavens were frequently childish, yet they were far from being unlettered men. Many had a breadth of knowledge that would have been the envy of later-day divines. Orosius, Isidore, Gregory, the Venerable Bede and Raban Maur wrote encyclopaedias

[1] Sarton: *op. cit.*, Vol. I, p. 17.

covering almost all branches of learning. Specialists were practically unknown.

The person to set this fashion of compiling encyclopaedias was Orosius, a Spanish priest of the fifth century. Apparently he was prompted by St. Augustine, with whom he lived for several years at Hippo, to supplement the *De Civitate Dei* by a history of the world. But it was history with a difference. Although his work was, in a sense, encyclopaedic its main concern was to show that Christianity was not responsible for recent calamities,[1] but that on the contrary, greater misfortunes had befallen the Empire in pagan times, and, so far from undermining it, the new religion had enhanced the strength of civilizing influences in every country where it had been embraced. The *Historia adversum paganos*,[2] as it was styled, is therefore a controversial document, and is radically unhistorical in method. However, its treatment is far better than its title. The introductory chapter on the countries and nations of the world is lucid, sober and fairly well informed. Orosius' reason for inserting a geographical introduction was that—to use his own words—' when the theatres of war and the ravages of disease shall be described, whoever wishes to do so may the more easily obtain a knowledge not only of the events and their dates but of their geography as well '. To a considerable extent it is independent both of Pliny and Ptolemy, being based upon earlier sources such as Strabo, though he nowhere cites his authorities. The following extracts are representative of Orosius' method and outlook.

Our elders made a threefold division of the world, which is surrounded on its periphery by the Ocean. Its three parts they named Asia, Europe and Africa. . . . Asia, surrounded on three sides by the Ocean, stretches across the whole East. Towards the West, on its right, it touches the border of Europe near the North Pole, but on its left it extends as far as Africa, except that near

[1] He was writing shortly after the fall of Rome to the Goths in A.D. 410.

[2] *Vide* Early English Texts Society, Vol. 79, 1883 (Translation by I. W. Raymond.).

Egypt and Syria it is bounded by Mare Nostrum which we commonly call the Great Sea.[1]

Europe begins, as I have said, in the north at the Tanais River (Don) where the Riphaean Mountains, standing back from the Sarmatian Sea, pour forth the Tanais flood. The Tanais sweeping past the altars and boundaries of Alexander the Great to the territories of the Rhobasci, swells the Palus Maeotis (Sea of Azov) whose immense overflow spreads afar into the Euxine Sea near Theodosia. From the Euxine near Constantinople a long narrow body of water leads to the sea which we call Mare Nostrum. The Western Ocean forms the boundary of Europe in Spain at the very point where the Pillars of Hercules stand near the Gades Islands and where the ocean tide comes into the Straits of the Tyrrhenian Sea.[2]

Africa begins with the land of Egypt and the city of Alexandria. On the shores of that great sea (Mediterranean) the walls of which touch all the continents and the lands in the centre of the earth, we find the city of Paraetonium (Marsa Labeit). From there the boundaries of Africa lead through districts which the inhabitants call Catabathmon (Akaba), not far from the camp of Alexander the Great above lake Chalearzus, whence they pass near the lands of the Upper Avasitae and across the deserts of Ethiopia to reach the Southern Ocean. The western boundary of Africa is the same as that of Europe, that is, the entrance of the Straits of Gades; its furthest boundaries are the Atlas Range and the islands which people call Fortunate. . . .

Much more land remains uncultivated and unexplored in Africa because of the heat of the sun than in Europe because of the intensity of the cold, for certainly almost all animals and plants adapt themselves more readily and easily to great cold than to great heat. There is an obvious reason why Africa, so far as contour and population are concerned, appears small in every respect (i.e. when compared with Europe and Asia). Owing to her natural location the continent has less space and owing to the bad climate she has more desert land. . . .[3]

Lower Egypt is bounded by Syria and Palestine on the East, by Libya on the West, by Mare Nostrum on the North and on the South by the mountain which is called *Climax*, together with Upper Egypt and the Nile. This river appears to issue forth

[1] *Historia*: I, 2. [2] *Ibid.*: I, 2. [3] *Ibid.*: I, 2.

from a spring near the mouth of the Red Sea at a place called the emporium of Mossylon.[1] Thence it flows towards the West for a long distance, forming in its midst the island called Meröe; finally bending to the north and swollen by seasonal floods, it waters the plains of Egypt. Some writers say that the Nile has its source not far from Mt. Atlas and then gradually disappears in the sand, from which after a brief interval, it flows out into a huge lake and then glides eastward through the Ethiopian Desert towards the Ocean (Indian), and finally, turning to the left, flows down to Egypt. Of a truth there is a large river of this kind which may have such an origin and course, and which, indeed, begets all the monsters of the Nile—although the barbarians living near its source call it Dara and the rest of the natives call it Muchul. However, this river is received and absorbed by a vast lake in the land of a people called the Libyo-Egyptians, not far from that other river which, as we have said, flows forth from the shores of the Red Sea—unless, as may be the case, it flows by a subterranean course into the bed of that river which flows down from the East.[2]

Upper Egypt stretches eastwards a long way; to the North is the Arabian Gulf; to the South, the Ocean; in the West (*sic*), Lower Egypt; in the East, the Red Sea. All told it contains twenty-four tribes.

On European countries, Orosius writes with fuller information. Thus:

Britain, an island in the Ocean, extends a long distance to the north: to its south are the Gauls. The city called Portus Rutupi[3] affords the nearest landing place for those who cross the water.

[1] Orosius' theory that the Nile takes its rise near the mouth of the Red Sea and the emporium of Mossylon and flows thence to the west as far as the island of Meröe, is not found in any earlier writing. 'It seems probable that this strange idea arose from a confusion between the *names* of Mossylon and the Massyli (or Massaesyli) where Juba, as reported by Pliny, supposed the Nile to rise for the second time (*Nat. Hist.:* V, 9, par. 52). In Solinus we find the promontory of Mossylon converted into Massylicum promontorium (cap. 56).'—Bunbury: *op. cit.*, Vol. II, p. 692.

[2] *Historia:* I, 2.

[3] i.e. Richborough in Kent.

From this point Britain faces directly the territory of the Menapi and Batavi,[1] which are located not far from the land of the Morini in the South. This island is 800 miles long and 200 miles wide. In the limitless ocean which stretches behind Britain are the Orcades Islands (Orkneys) of which twenty are deserted and thirteen inhabited. . . . Ireland is quite close to Britain and smaller in area. It is, however, richer, on account of the favourable character of its climate and soil. . . .

For all its limitations, the *Historia* proved to be very popular during the centuries that followed. As the first continuous history of the world written from a Christian angle, many students preferred it to world histories written by pagans. To study history was to read Orosius and he was quoted by almost every Christian encyclopaedist down to about A.D. 1300. Citation of passages does not, of course, prove that the entire work was read: at times the passages from Orosius were without doubt quoted from other works. Among those who made use of the whole text were Alfred the Great and Isidore of Seville. But even as late as *c.* 1410, we find Pierre d'Ailly referring to it with the utmost respect. Nevertheless, Orosius was soon entirely eclipsed by Isidore of Seville, a bishop of the seventh century (*fl. c.* A.D. 600–636) and the author of the most representative compilation of the Middle Ages. This work—entitled *Origins* or *Etymologiae*[2]—comprises twenty books, two of which, the thirteenth and fourteenth, deal with geographical topics. The thirteenth deals with the world as a whole, with the Ocean, the seas both open and enclosed, the tides, rivers, winds, &c., i.e. with physical geography. In the fourteenth book Isidore enumerates and briefly describes the political divisions of the earth. In view

[1] *Vide infra*, chap. iv.

[2] So called because the author always gives the etymology of everything he describes or defines. Indeed, the tenth book contains only the etymological definition of words alphabetically arranged. Isidore wrote a number of other works (N.B. *De Natura Rerum*), but the *Etymologiae* contain in condensed form nearly everything written elsewhere. The most accessible edition of the *Etymologiae* is the one edited by W. M. Lindsay in the *Oxford Classical Texts*, 2 vols., 1911.

of the extraordinary influence which this book wielded we shall quote from it at some length.

Concerning the earth we are told that it is 'named from its roundness (*orbis*) which is like a wheel; whence a small wheel is called *orbiculus*. For the Ocean flows round it on all sides and encircles its boundaries.'[1] As to size, Isidore accepts the Macrobian estimate of 252,000 stadia, but by employing the more usual reckoning of eight stadia[2] to the mile and 87½ miles to the degree, obtains the grossly exaggerated figure of 31,500 miles for the circumference of the earth. As usual the earth is divided into three parts: Asia, Europe and Africa.

The ancients did not divide these three parts of the world equally, for Asia stretches right from the south, through the east to the north, but Europe stretches from the north to the west and thence Africa from the west to the south.[3] From this it is quite evident that the two parts Europe and Africa occupy half the world and that Asia alone occupies the other half. The former were made into two parts because the Great Sea (called the Mediterranean) enters from the Ocean between them and cuts them apart. . . .[4]

Proceeding to a systematic description of the countries of the world, Isidore says of Asia that it is

bounded in the east by the sunrise, on the south by the Ocean, in the west by the Mediterranean, in the north by Lake Maeotis (Sea of Azov) and the river Tanais (Don). It contains many provinces and districts whose names and geographical situations I will briefly describe, beginning from Paradise. . . .[5]

Paradise is a place lying in the eastern parts, whose name is translated out of the Greek into Latin as *hortus* (i.e. garden). It is called in the Hebrew tongue Eden, which is translated in our language as *Deliciae* (i.e. place of luxury or delight). Uniting these two gives us Garden of Delight; for it is planted with every kind of wood and fruit-bearing tree, having also the tree of life.

[1] XIV, 2, i.

[2] 1 stadium equalled 625 feet in Isidore's estimation.

[3] *Etymologiae*: XIV, 2, ii.

[4] *Ibid.*: XIV, 2, iii.

[5] *Ibid.*: XIV, 3, i.

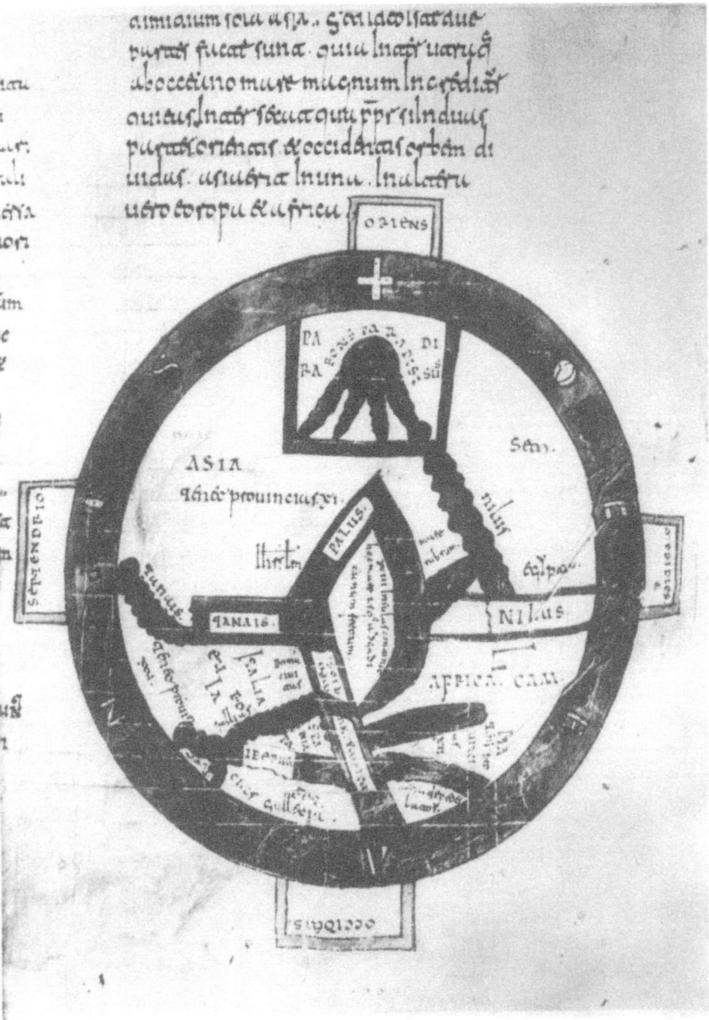

III

WORLD MAP FROM TENTH CENTURY MS. COPY OF ISIDORE'S *Etymologiae*
(*Academia de la Historia, Madrid*)

There is neither cold nor heat there but a continual spring temperature.[1]

From the middle of the Garden, a spring gushes forth to water the whole grove and, dividing up, it provides the sources of four rivers.[2] Approach to this place was barred to man after his sin, for now it is hedged about on all sides by a sword-like flame (*romphæa flamma*), that is to say it is surrounded by a wall of fire that reaches almost to the sky.[3]

This strong biblical note coming so early in the topographical section of the work might lead us to expect its continuance in subsequent chapters ; but apart from one or two entirely understandable references to biblical lore—'Scythia and Gothia also are said to have been named by Magog son of Japheth '[4] and ' the River Ganges which the sacred scripture calls Phison, flows down from Paradise to the realms of India '[5]—only the most sparing use is made of this source of information. By far the greatest percentage of Isidore's material is culled from pagan sources ; indeed, much of his geography might have been written by late classical writers such as Mela and Solinus, to whose limitations and errors he is, perforce, heir. His treatment of the habitable earth enables us to arrive fairly easily at the scope of his knowledge. In the extreme east of Asia the country of Seres ' rich in fine leaves, from which are cut fleeces which the natives who decline the merchandise of other peoples sell for use as garments '[6] bounds his horizon ; beyond there is only the Scythian ocean flowing from the ' Caspian to the Eastern Sea '.[7] To northward lies Scythia, stretching from the ' Seric ' (i.e. Eastern) ocean in the east to the Caspian Sea in the West. ' Several of the districts are rich, but some are uninhabited, for while they are rich in gold and precious stones they are rarely approached by man owing to the ferocity of the

[1] *Etymologiae :* XIV, 3, ii.
[2] *Vide*, Plate No. III.
[3] *Etymologiae :* XIV, 3, iii.
[4] *Ibid. :* XIV, 3, xxxi. [5] *Ibid. :* XIII, 21, viii.
[6] *Ibid. :* XIV, 3, xxix. [7] *Ibid :* XIV, 3, xxix.

Griffens.'[1] The land of Hyrcania, bordering Scythia to the west, 'has many tribes wandering far afield on account of the unfruitfulness of their lands.'[2]

Europe, in true classical fashion, is divided from Asia by the River Don and is bordered on the North by the Northern ocean. Hard by it, and forming the *ne plus ultra* of the known world, is the land of "Barbaria"—so called on account of the wild tribes inhabiting it.[3] Enumerated among these are the Alani, the Dacians, the Goths and the Suevi. 'Thyle is the furthest island in the Ocean in the Northern and Western waters beyond Britain.'[4] According to Isidore, 'it takes its name from the sun, because here the sun makes its summer solstice and beyond the summer day is of great length'.[5] The sea around is 'calm and frozen'—but presumably not in summer.[6] The westward limit of the world is furnished by the Fortunate Isles, so named because

> they are blessed with abundance of produce; their woods yield apples naturally, their ranges of hills are clad with unplanted vines and everywhere there are crops and vegetables in place of pasture. Hence the false opinion of pagans, and the poems of secular poets, claiming that these islands were Paradise. They are situated in the Ocean off the coast of Mauretania.[7]

Concerning Africa our author says little that enables us to

[1] *Etymologiae*: XIV, 3, xxxii. In Book XII of the *Etymologiae*—on animals—Isidore describes them thus: 'The Gryphes are so called because they are winged quadrupeds. This kind of wild beast is found in the Hyperborean Mts. In every part of their body they are lions, and in wings and head are like eagles, and they are fierce enemies of horses. Moreover they tear men to pieces.' Hesiod seems to have been the first writer to mention them: but it was Herodotus and later Pliny who did so much to popularize belief in them.

[2] *Ibid.*: XIV, 3, xxxi, an early reference to the nomadism of the Asiatic steppes.

[3] *Ibid.*: XIV, 4, iii. [4] *Ibid.*: XIII, 6, iv. [5] *Ibid.*

[6] '. . . having its name from the sun, because there the sun makes its summer halt, and there is no day beyond it: whence the sea there is sluggish and frozen'.

[7] *Ibid.*: XIII, 6, viii.

put bounds to it; he only tells us that 'it begins at the boundaries of Egypt,[1] continuing to the South through Ethiopia to Mount Atlas'.[2] As to Ethiopia, he avers that

> the whole of it is under the southern pole (i.e. hemisphere). Towards the west it is mountainous; in the middle it is sandy; to the east a desert. . . . In the south it is bounded by the Ocean, and in the north by the River Nile. It contains very many tribes of different aspects, with strange forbidding countenances.[3]

The south-eastern horizons of the world are circumscribed by the coasts of India—'containing many tribes and towns'—the island of Taprobane,[4] Chryse,[5] Argyra,[6] and Tyle, 'which is never without leaves on its trees'.[7] In connexion with Isidore's account of Taprobane, we are treated to some fictitious statistics, for he tells us that it stretches

> 875 miles in length and 625 in width. It is separated (from India) by a river that flows between. It is rich in pearls and precious stones; part of it is, however, infested with wild animals, but part is occupied by men. In this island they say that there are two summers and two winters in one year and that the flowers bloom twice.[8]

We have quoted the *Origins* at sufficient length to disclose the kind of geographical knowledge Isidore possessed and that was current presumably round about his time. It is cast in an antiquarian mould and contains nothing that cannot be traced back, either directly or indirectly, to popular written Latin sources.[9]

[1] Egypt is regarded as part of Asia.

[2] *Etymologiae:* XIV, 5, ii.

[3] *Ibid.:* XIV, 5, xiv.

[4] i.e. Ceylon.

[5] i.e. The Golden Chersonese (Malay Peninsula?) spoken of by Ptolemy, *vide Geography*, VII, 4.

[6] Cattigara?

[7] *Ibid.:* XIV, 3, v.

[8] *Ibid.:* XIII, 6, xii.

[9] It seems probable that his working library contained works of the following authors: Lactantius, Tertullian, Jerome, Ambrose, Augustine, Orosius, Cassiodorus, Suetonius, Pliny, Solinus, Hyginus, Sallust, Hegesippus, the abridger of Vitruvius, Servius, the scholia on Lucan and Justinus. *Vide* E. Brehaut: *An Encyclopaedist of the Dark Ages*, p. 47.

Occasionally whole paragraphs have simply been extracted from these sources and employed as though they were the author's own.[1] In at least one instance a passage has been wrenched from its context and given an entirely new setting. This is where the inhabitants of Ethiopia are described in exactly the same words as Solinus uses when speaking of the Indians.[2] Had he lived in modern times Isidore would have suffered badly under the Copyright Act—but, then, so would they all! For his chief authorities he goes to the late and post-classical redactors, notably Solinus and Orosius, whose own works are themselves little more than compilations. In his *Quaestiones in vetus Testamentum*, he confesses his indebtedness by admitting that 'my voice is but their tongue'. At the same time his work is quite remarkable for its breadth of learning, embracing as it does all the liberal Arts and such subjects as Natural History, Agriculture, Architecture and Military Science, and may be taken as representing the widest possible scope in secular knowledge that an orthodox Christian of the Dark Ages could allow himself. 'Indeed so hospitable an attitude toward profane learning as Isidore displayed was unparalleled in his own period, and was never surpassed throughout the Middle Ages.'[3]

The *Origins*, therefore, like so many medieval treatises, is a compilation from compilations. In some ways this is all to the good, for it gives Isidore a representative quality which a more original writer would have lacked. As it is, his work provides 'a cross section of the mind of the Dark Ages'[4] and a summary of what was known (or perhaps of what Isidore thought ought to be known) at a time when the de-secularizing process, which European thought had been undergoing since the early days of the Christian era, was almost complete.

[1] cf. *Etymologiae*: XIV, 3, xxix, xxxii, with Solinus, *Collectanea*: . . . chaps. 20, 33.
cf. *Etymologiae*: XIV, 4, iii, with Orosius, *Historia*: . . . I, 2.
[2] *Etymologiae*: XIV, 5, xvi.
[3] E. Brehaut: *An Encyclopaedist of the Dark Ages*, p. 31.
[4] C. H. Haskins: *The Renaissance of the 12th Century*, p. 304.

Furthermore, whereas many of the Encyclopaedists enjoyed only an ephemeral popularity, Isidore remained in high esteem down to the fourteenth and even the fifteenth centuries ; his influence was such that

> whenever you have a geographical treatise or a section on geography in a larger work, you are sure to find Isidore is one of the most important sources. Several anonymous poems and treatises of the eighth and ninth centuries depend mainly on him, and in the *Liber Glossarum*, a huge encyclopaedia-cum-dictionary compiled early in the eighth century, the geographical sections are culled primarily from that author.[1]

The cause of this long-continued favour is somewhat obscure, for it cannot be seriously argued that Isidore's work was intrinsically superior to others of the same period. H. O. Taylor, showing little insight, explains it on the ground of 'his unerring faculty of selecting for his compilations the foolish and the flat' ![2]

Rather less orthodox than Isidore was the anonymous Ravennese Geographer of the mid-seventh century. For devout Christian that he was, he had no compunction about using the works of Porphyry, Iamblichus, Aristarchus and Lollianus, all pagan philosophers. Indeed, most of his topographical material was garnered from non-Christian sources : prominent among them were the Imperial road itineraries (including the Peutinger Table), the works of Jordanis (i.e. Jornandes) on the history of the Goths, Castorius (unknown in history, but presumably a Roman cosmographer of the third century) and even Ptolemy, whom he styles, erroneously, of course, 'King of the Macedonians in Egypt'. The balance is redressed, however, by an introductory rhapsody upon the greatness and glories of God's creation, the occasional use of Holy Writ to bolster up some tottering pro-

[1] M. L. W. Laistner : *Travel and Travellers in the Middle Ages*, p. 37 (Edited by A. P. Newton.). Much the same thing is true of Lambert of St. Omer's *Liber Floridus* (*c.* A.D. 1120).

[2] *The Mediaeval Mind :* Vol. I, p. 89.

position, and numerous citations from Orosius, Rigilinus, Isidore and the early Fathers. From his *Cosmographia*, we learn that the Ravennese Geographer regarded the earth's figure as approximately round, and surrounded by ocean, but the ocean was not entirely continuous for it did not extend behind India, which was the Eastern limit of his *oikoumenē*. Beyond India lay an impassable desert. To attempt to pry into the secrets of those Eastern boundaries was flat blasphemy for Christians, for does not Scripture say that no mortal man can penetrate to the hidden Paradise of God, which is in the farthest East? As to the Northern parts of the earth, he says it is clear from the sayings of philosophers, that, beyond the ocean, there are mighty mountains placed by command of God. These make night and day (as Cosmas also maintained) by forming the screen behind which sun and moon disappear during the circuit of their course. Some, apparently, denied the existence of these mountains, declaring the Ocean to be the end of the inhabited world. Impudently they ask 'Whoever saw these mountains with his own eyes? Where are they named in scripture?' To which he replies: 'Where indeed? Is it not clear that although known to the Creator they have been forbidden by Him to human knowledge and are therefore inaccessible?' As to the scriptural difficulty, is there not a plain reference in the Book of *Genesis*, he asks? 'The sun came up over the land when Lot entered into Zoar'.[1] Does not the 'coming up' of the sun presuppose a height to be surmounted? Lest his arguments should still be attacked by the sceptics, he hastens to add 'but the manner and cause of this are known only unto God'.[2]

Although the works of Isidore and the anonymous Ravennese have a really representative character, it would be a serious mistake to suppose that they typify the spirit or the scholarship of a period so long and inchoate as the Dark Ages. They are typical of that school of thought—and it had a big following—which recognized no fundamental opposition between science and religion and which did, quite unhesitat-

[1] *Genesis* xix. 23. [2] *Cosmographia*: I, 2–9.

ingly, temper its orthodoxy with pagan culture. To men of this conviction, the Earthly Paradise—' a country in the farthest East, beyond all known land '[1]—was just as much an established fact as elephants in India. But many scholars had a stronger ecclesiastical bias and regarded the scriptures as carrying an *imprimatur* altogether more compelling than that of classical literature. They ranged from ' moderates ' like Raban Maur and Basil to ' extremists ' like Cosmas. Here, for instance, is what Maur,[2] in his encyclopaedia entitled *De Universo* says about the figure of the earth.

The world is so called from its roundness (*orbis*) which is like a wheel. Whence a small wheel is called *orbiculus*. For the ocean flows round it on all sides and encircles its boundaries.[3] The *world* (*orbis*) signifies either the whole world of history or the universal church of allegory ; it is also used, in the opposite sense, to describe the impious who are often called by the name of the *world*. About these the Psalmist says : ' He shall judge the world with righteousness.'[4] They (the people of the world) are to be judged by the Lord, as by reason of their frowardness they are without remission. . . . It is well that we should inquire what the Psalmist means by ' the circle of the earth ' when he says that ' His lightnings lightened the world ' (*orbis terrarum*)[5] and why, in several other places, he says that the earth is comprised of the same figure. On the other hand, however, in the 106th[6] Psalm he comprises the earth's expanse under four cardinal points, saying : ' From the east and the west, from the north and from the south '. A very similar statement occurs in the Gospel—where it says : ' He shall send his angels with a great sound of a

[1] *Cosmographia* (Ravenna) : I, 5.

[2] Maur was born at Mainz *c.* 776. He was educated at the Abbey of Fulda and later at Tours under Alcuin. In 804, after the death of Alcuin, he was made principal of the monastery school at Fulda, and later, in 822, Abbot. He resigned twenty years later to devote himself to literary labours. He died in 856. His works, mainly commentaries, occupy six volumes of Migne's *Patrologiae Cursus Completus, sive Bibliotheca . . . series Latina.*

[3] cf. *Etymologiae :* XIV, 2, i., *ante*, p. 24.

[4] *Psalm* xcvi. 13.

[5] *Psalm* lxxvii. 18.

[6] Vulgate numbering : in our version *Psalm* cvii. 3.

trumpet and they shall gather together [his elect] from the four corners of the earth.[1] Whence it is fitting, I think, to inquire how far the quadrate and circular shapes of the earth can agree, when the figures themselves, as geometricians maintain, are different. The Scriptures call the shape of the earth a circle for this reason, because to those who look at its extremity (i.e. round the horizon) it always appears as a circle. This circle the Greeks call the *horizon*,[2] signifying that it is formed by the four cardinal points (or hinges, pivots) ; these four points signify the four corners of a square which are contained within the aforesaid circle of the earth. For if you draw two straight lines from the East, one to the South and one to the North, and in the same way also draw two straight lines from the western point, one to each of the aforesaid points, namely the south and the north, you make a square of earth within the aforesaid circle. How this aforesaid square (*demonstrativus quadratus*) ought to be inscribed within the circle, Euclid clearly shows in the Fourth Book of the Elements. Wherefore the Sacred Scripture rightly calls the shape of the earth circular, and says that it is comprehended by four cardinal points.[3]

From here onwards, Maur continues his chapter, as he had begun, with words identical with those of the *Origins*. However, when he arrives at the tripartite division of the habitable earth, he interpolates the following biblical gloss :

And most appropriate is this division of the earth into three parts, for it has been endowed with faith in the Holy Trinity and instructed by the Gospels, where we read the words of the Saviour that the world [4] is like unto leaven which a woman took and hid in three measures of meal [5] until the whole was leavened.[6] That is to say, the Holy Church on earth—which has been peopled from the three sons of Noah—leavened by the doctrine of the Gospel, will penetrate into the hearts of the faithful, until the whole race of mankind be converted by righteousness and spiritual knowledge to the worship and service of God.[7]

[1] N.T. 'four winds', *vide Matthew* xxiv. 31.
[2] i.e. bounding, ὁρίζων in Greek.
[3] *De Universo :* XII, 2.
[4] N.T. 'kingdom of heaven'.
[5] The three continents.
[6] *Matthew* xiii. 33.
[7] *De Universo :* XII, 2.

Further reading only serves to show that the entire geographical section of the *De Universo* is composed of passages lifted bodily from Isidore's *Origins* (or from a common source), interwoven with biblical commentaries. The difference, accordingly, between this ninth-century compilation and its seventh-century predecessor is largely one of emphasis and not of knowledge. Raban Maur merely changed the arrangement and devoted a larger part of his parchment to religious topics. He added further material gleaned from the Church Fathers. This further matter consisted of 'the mystical interpretation of things',[1] which he subjoined to their 'natural' explanations. Just as the medieval ecclesiastic believed that there was a higher as well as a literal interpretation of a given passage of Scripture, so the medieval scientist came to believe that even material things had higher meanings, that the things which were visible and physical were but types and shadows of the things which were invisible and spiritual.[2] How far Maur succeeds in establishing such a thesis, the reader may judge from the foregoing exegesis.

Basil adopts a somewhat more conservative standpoint in his *Hexaemeron*—a treatise, as the name suggests, on the six days' work of Creation.[3] Taking as his initial premise the inerrancy of the Biblical cosmogony, he finds himself in frequent conflict with Greek learned opinion. On these occasions, he usually prefers to be guided by 'the simplicity of faith than the demonstrations of reason'.[4] Even so, he is not entirely above qualifying a Biblical statement when it suits his purpose. Thus, he interprets the command 'Let the waters under heaven be gathered together into one place' to apply only to the sea or ocean, which he contends is one body of water, and not to pools and lakes, recognizing that otherwise 'our explanation of the creation of the world may appear contrary to experience, because it is evident that all the waters do not flow together in one place'. In this connexion he states that 'although some

[1] *De Universo*: Preface.

[2] cf. the 'archetypes' of Plato and Socrates.

[3] *c.* A.D. 360–380.

[4] *Homily*: X.

authorities think that the Hyrcanian and Caspian Seas [1] are enclosed in their own boundaries, if we are to believe the geographers, they communicate with each other and together discharge themselves into the Great Sea '[2] Where no such conflict of loyalties is at issue, he is quite happy to accept the findings of pagan philosophers and scientists, drawing freely upon the works of Aristotle (*Meteorology*, &c.) and Pliny.

At the other extreme stands a man like Cosmas Indicopleustes (i.e. Indian traveller), who despised all trafficking with pagan lore. For him the Scriptures were adequate and profitable not only for doctrine, reproof and instruction in righteousness, but for instruction in earth-knowledge as well. His *Christian Topography*, written in the sixth century, seems to the modern reader neither topographical nor Christian, but merely a crude caricature of cosmography. The following are some typical passages :

Firstly, his view of those who say that the earth is round :

With supercilious air, as if they surpassed in wisdom the rest of mankind, they attribute to the heavens a spherical figure and a circular motion and by geometrical methods and calculations applied to the heavenly bodies, as well as by the abuse of words and by worldly craft, endeavour to grasp the position and figure of the world by means of the solar and lunar eclipses, leading others into error, while they are in error themselves in maintaining that such phenomena could not represent themselves if the figure was other than spherical.[3]

Subsequently, Cosmas gives us his own personal view of the shape of the earth in the following, original terms :

The Deity . . . having founded the earth, which is oblong, upon its own stability,[4] bound together the extremities of the

[1] In classical geography the Hyrcanian and Caspian Seas were one and the same. Perhaps Basil is confusing the Hyrcanian Sea with the Aral Sea, for Hyrcania was a province of the ancient Persian Empire situated to the south of the Caspian Sea.

[2] *Homily* : IV, 2–4.

[3] Edited by J. W. McCrindle (Hakluyt Society, Old Series, Vol. 98), p. 8.

[4] *Vide Psalm* cii. 5, *Job* xxxviii. 4–6.

> heaven with the extremities of the earth, making the nether extremities of the heavens rest upon the four extremities of the earth, while on high he formed it into a most lofty vault overspanning the length of the earth. . . . The figure of the earth is lengthwise from east to west, and breadthwise from north to south and it is divided into two parts ; this part which we inhabit and which is all encircled by the intermedial sea, called the Ocean by the pagans, and that which encircles the Ocean and has its extremities bound together with those of heaven . . .[1]

On the other burning question of 'Christian topography', namely the existence of an Antipodal landmass, Cosmas is as eloquent as he is sarcastic. To him the whole thing is nothing more than an

> old wives' fable. For, if men, on opposite sides, placed the soles of their feet each against each, whether they chose to stand on earth or water, or air or any kind of body, how could both be found standing upright ? The one would assuredly be found in the natural upright position and the other, contrary to nature, head downwards. Such notions are opposed to reason and alien to our nature and condition.[2]

With Cosmas and his kind the theological argument against a spherical earth carried great weight also. In a sentence it was this :

The apostles were commanded to go into all the world and preach the Gospel to every creature : they did not go to any such part of the world as the Antipodes : they did not preach to any creatures there : *ergo*, no Antipodes exist.

But Cosmas and the other supporters of the flat earth theory did not have it all their own way—even in the Dark Ages. Indeed, it has been claimed that there is at least one witness in every century to the tradition of the sphericity of the earth during this epoch and that these witnesses include such

[1] *Op. cit.* p. 30.
[2] McCrindle, *op. cit.*, p. 17. Cf. *Lactantius : Divinae Institutiones* III, 34.

men as Hilary of Poitiers, Ambrose of Milan, Augustine, Cassiodorus and Isidore of Seville.[1] Unfortunately for the force of the contention none of these men makes an unambiguous affirmation of belief that the earth is a globe. On the contrary, the relevant passages of their works admit, in some cases, of a construction not incompatible with the flat earth hypothesis. The case of Isidore is particularly apposite. Betten avers that Isidore follows Ptolemy.[2] True, he quotes writers of Antiquity who favoured a round earth —he even uses the word *sphera* [3]—but he also makes statements, apparently to be taken as his personal view, which can be reconciled only with a belief in a flat earth. Thus, in the third book of the *Origins* we read: 'The size of the sun is greater than that of the earth and so, from the moment when it rises, it appears equally to east and west at the same time.'[4] Elsewhere, it is fairly clear that a sphere and a circle meant very much the same thing to Isidore.[5] And so the suspicion is aroused that when he is stating astronomical ideas he is copying something he does not understand. Such a suspicion receives a striking confirmation where he is dealing with the theory of terrestrial zones (which he had borrowed from Hyginus). Here, in *De Natura Rerum*,[6] he speaks of the northern and southern circles (both of which he regards as uninhabitable from cold) as 'being adjacent to each other' and so he shows them to be in an accompanying figure. From this [7] it is evident that he thought of zones and circles as interchangeable terms and that they did not run round the circumference of a spherical earth, but lay flat on a flat earth where they filled the *orbis terrae* or circle of the earth.

But while this may dispose of one of Betten's chief witnesses, there can be no gainsaying the essential truth of his contention. As we have already seen, even so strong an advocate of the flatness of the earth as Cosmas had to admit the existence of

[1] F. S. Betten: *Catholic Historical Review*, Vol. 3, 1923, pp. 74–90.
[2] *op. cit.*, p. 84. [3] *De Natura Rerum*: cap. 48. [4] *Ibid.*: cap. 47.
[5] *Vide De Natura Rerum*, cap. 12, and *Etymologiae*, XIV, 2.
[6] Cap. 10. [7] *Vide Plate* No. IV.

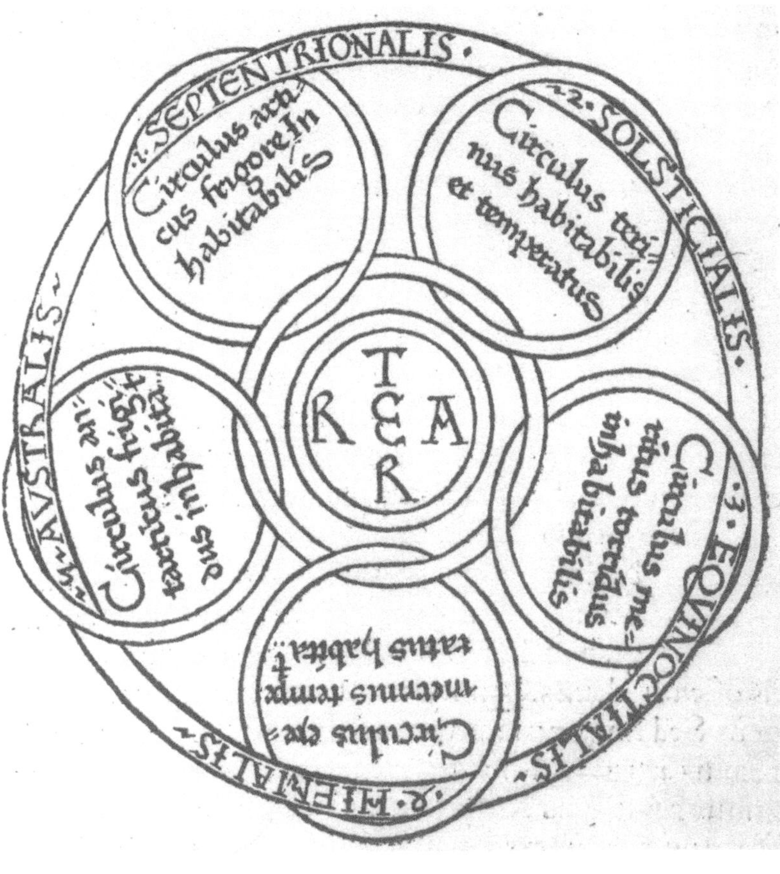

IV

DIAGRAM OF ZONES FROM ISIDORE'S *Etymologiae* (1472)

men who were of a contrary persuasion. Lactantius even condescends to give us their arguments:

> If you inquire [says he] from those who defend these marvellous fictions why all things do not fall into that lower part of the heaven, they reply that such is the nature of things that heavy bodies are borne to the middle, and that they are all joined together towards the middle, as we see spokes in a wheel: but that the bodies that are light as mist, smoke and fire, are borne away from the middle, so as to seek the heaven.[1]

In seeking further support for the round-earth protagonists, we must beware of arguing that belief in the existence of the Antipodes is, *ipso facto*, evidence of belief in the rotundity of the earth, for during the Middle Ages the Antipodes, in the exact sense of that term, became confused with a fourth, or austral continent uninhabited and lying beyond the impassable equatorial ocean.[2]

By the eighth century the Church appears to have largely forgotten its early doubts about the shape of the earth and to have accepted the saner opinions of the Ancients.[3] In

[1] *Divinae Institutiones*: III, 34.

[2] e.g. Maur: *De Universo*, VIII, 4. The reasons why no orthodox persons were prepared to allow the possibility of human dwellers in this region were twofold; first, on 'rational' grounds—as in the case of Cosmas: second, on 'Scriptural' grounds. 'No one', says St. Augustine, 'will deny that these (i.e. the members of the human race) are all descended from that one man (i.e. Adam) so that all the races which are reported to have diverged in bodily appearance from the usual course which Nature generically and almost universally preserves, . . . unquestionably trace their pedigree to that one first father of all. . . .' Continuing, he declares: 'It is too absurd to say that some men might have taken ship and traversed the whole wide ocean and crossed from this side of the world to the other, and that, thus, even the inhabitants of that distant region are descended from that first man' (*De Civitate Dei*: XVI, 8). And as the Church could not brook the idea of a polygenetic humanity, if only because in their view it implied that Christ must have suffered and died more than once, the Antipodes remained unoccupied. At the same time, rumours were continually being put about that 'within its confines a race of Antipodeans is fabulously said to exist' (*Isidore*: *Etymologiae*, XIV, 5, xvii).

[3] *Vide* Bede: *De Natura Rerum*, cap. 46.

view of the lively interest of Christian scholars in classical learning at this epoch, this is not so surprising as it appears. For instance, Bede tells us that the first Abbot of Wearmouth, Benedict Biscop, went seven times, including five times from Britain, in pious pilgrimage to Rome 'and brought back many books of all subjects of divine learning, which had been bought at a price or been given him freely of his friends '.[1] In another place we discover that among these books was 'a manuscript of the cosmographers . . . most excellent for workmanship '.[2] About the same time, up and down the continent, and in the town of Armagh in Ireland, monastic schools [3] were being established where the Greek language, lost awhile, was once more being taught. These and parallel movements, which constituted the so-called 'Carolingian Renaissance,' left their mark on the literature of the time. Thus it is not possible to read the writings of Alcuin, the head of the School and Library at York and, later, the leading spirit in the revival in Charlemagne's realm (*c.* 782), or those of the great ninth-century Irish thinker, Scot Eriugena, without realizing that they are of a different intellectual calibre from those of their predecessors. Notwithstanding a timid mistrust of pagan learning, Alcuin loved the temple of the Muses, and was at once their high-priest and apostle in the days when the worshippers at their shrine were few. At the same time, he himself seems to have been a devout scholar concerned, above all else, to put the fortunes of learning—classical as well as ecclesiastical—in a position of advantage, safeguarded by an abundance of truthfully transcribed books, sheltered within the Church and defended by the civil power.[4] As for Eriugena, his work evinces a spirit both critical and constructive and a breadth of sympathy unparalleled in its day. His main work, *De Divisione Naturae*,

[1] *Lives of the Abbots* (Historical Works of the Venerable Bede, published in Loeb Classical Library), pp. 399–401.
[2] *Ibid.* : p. 429.
[3] e.g. at Ravenna, Rome, Fulda.
[4] *Vide* A. F. West : *Alcuin*, p. 122 *et seq.*

the leading idea of which is the unity of nature proceeding from God and ultimately resolved into Him, draws upon a wider field than was customary at that time. For instance, it shows more than a passing acquaintance with the writings of Eratosthenes, Pliny, Ptolemy (he even speaks of his *Geography*), Capella, besides the usual patristic authorities such as Basil, Chrysostom, Ambrose, Jerome and Augustine. But this does not reduce the work to the level of a mere compilation, for although most of the ideas were borrowed, the form and the system were his own, and this was no mean achievement.

> It was only a metaphysician of great insight who could have found all that he needed in the incomplete and disorderly literature to which he had access, and it was only a philosopher of great speculative power who could have wrought it all into so massive a system.[1]

Although the *De Divisione Naturae* runs to some quarter of a million words, Eriugena devoted only two or three chapters —in his third book—to the natural sciences of his day and for these he is largely dependent on Pliny's *Natural History*.

This growing intellectual vitality is reflected, if only faintly and temporarily, in the geography of these centuries. At the beginning of the eighth century, the Venerable Bede is writing about the recent travels of Arculf, the Gallican Bishop, and advances his own views about the tides.[2] Less than a century later (*c.* A.D. 825), Dicuil, the learned Irish monk, is describing the almost contemporary discovery of Iceland and the fresh-water canal between the Nile and the Red Sea,[3] while in a little-known astronomical treatise written *c.* A.D. 814 he criticizes the current planetary theories and remarks that if any one would give him a better solution

[1] H. Bett: *Johannes Scotus Erigena . . .*, p. 165.

[2] In his *De Temporum Ratione*, the most mature of his scientific studies, we find the first mention of the *port* or time-lag between the moon's meridian and high-water following, the interval being different from place to place—a statement based presumably upon his own observation.

[3] *De Mensura Orbis Terrae*: VI, 3 and VII, 2.

of the problem, he would gladly adopt it. Later in the same century, Alfred the Great translates the *Pastoral Care* of Gregory, the *De Consolatione Philosophiae* of Boethius and the *History* of Orosius into the vernacular, and adds his own reflections upon them and other matter not in the originals. In rendering Orosius, he re-wrote the geographical introduction, inserted a description of Germany (by which he understood all the country from the River Don to the Rhine and from the Danube to the Arctic Sea) and two accounts given him by his Norse liegemen, Ohthere and Wulfstan. The alertness of his mind is shown by the employment of the latest geographical knowledge. The description of Germany, although little more than a list of boundaries was, from the comparative accuracy of its topography, a real contribution to the knowledge of the time, drawn, it may well be, from the reports of travellers visiting the West Saxon Court. The tales of Ohthere, who doubled the North Cape, and of Wulfstan, who sailed from Schleswig to the eastern shores of the Baltic, certainly came from this source.

Now we shall look in vain through the works of Isidore and Orosius either for gleanings of contemporary history or for personal observations on natural phenomena. Nor shall we find in them any suggestion that their authorities were in need of revision. Such admissions, whether explicit or implicit, are by no means unknown in these later writings, however. Thus, in translating Orosius' *History* into Anglo-Saxon, Alfred did not hesitate to cut down the number of chapters by half and to expurgate and annotate the historical portions in accordance with those educational aims which were never far from his thoughts. Dicuil, on his part, is much concerned about the authenticity of certain stories reported by Solinus. In recording the latter's assertion that 'the snow summit of Atlas lies hidden above the clouds', he asks, more satirically than sensibly it is true, 'How could the offspring (i.e. the snow) lie above its source? Snow, falling as it does from the clouds of heaven, would indeed be a fine sight up above the self-same clouds '.[1] As for Bede, he finds himself compelled to forsake the flat-earth postulates of his predecessors for the spherical view ; while his argument is based primarily upon Pliny's *Natural History*, it has a vocabulary and an argumentation quite its own.[2]

But these glimmerings of insight and independent thinking are all too fugitive and do very little to dispel the intellectual gloom of the age. For the most part Dicuil and his contemporaries are quite content to draw upon the heritage of the past without comment or modification, and, cheek by jowl with sober statements on current happenings, make such extraordinary assertions as the following : that Mount Pelion is '250 miles from base to summit', that the Alps are only a modest 50 miles in height and that the whole world comprises 2 seas, 72 islands, 40 mountains, 65 countries, 281 towns, 55 rivers and 116 peoples.[3] Bede is not so prone to seek safety in numbers, but when he does, as in his division of the earth into different kinds of zones, he relies on the established traditions of Macrobius, Isidore and Pliny, with the result that these discussions are mere reversions to type. Thus :

> The earth is divided up into five zones, the differences between them being that certain temperate parts are inhabited, while some remain uninhabited on account of the rigour of their cold and heat. The first is the northern zone, uninhabitable from cold, whose stars never set for us. The second is the solstitial or summer zone from our point of view, and is temperate and habitable. The third is the equinoctial covered by the middle of the circle of the zodiac, torrid and uninhabitable. The fourth is the brumal or winter zone, on the lower side of the zodiac towards the southern pole, and is temperate and habitable. The fifth is the southern,

[1] *De Mensura Orbis Terrae :* IX, 3. Cf. Homeric conception of Mt. Olympus, *Odyssey*, VI, 41 *et seq.*

[2] *Vide De Natura Rerum*, cap. 46. This proved to be a very popular text-book in later centuries and was translated into Anglo-Saxon in the tenth century, if not before. *Vide* T. Wright : *Popular Treatises on Science written during the Middle Ages.*

[3] *De Mensura Orbis Terrae :* VIII, 26–31.

around the southern pole and is covered by land, but is uninhabitable by reason of the cold. The three middle zones are distinguishable by the inequality of their seasons, since the sun holds the one at the summer solstice while the others are at the equinox and at the winter solstice. The extreme zones are always without sun. Wherefore, from the island of Thule one day's sailing to the North, the frozen sea is reached.[1]

Bede in his territorial division is similarly beholden to current tradition, for it is an almost exact copy of Isidore's text. Such departures as he allows himself are not sufficiently significant to warrant quotation. The same thing is true, *mutatis mutandis*, concerning his classification of the Northern Hemisphere into ten zones—based upon the length of day—for it is nothing more than a verbal transcription of the last chapter of Book VI of Pliny's *Natural History*.

We see then, that apart from a somewhat more inquisitive and experimental temper, the geography of the Carolingian Age is much the same as that of the earlier centuries. In each case it is 'historical geography'. The more encyclopaedias we peruse, the more general we shall find this state of affairs to be. The text-books of this age, accordingly, bear little or no relation to the times in which they were composed; neither Macrobius, Orosius, Isidore, Maur, Bede nor Dicuil described the known world of his own day. The geography of the fourth century is essentially the geography of the eighth and is, indeed, very little different from that of the fifteenth, for Columbus (as Alexander von Humboldt first observed),[2] drew many of his ideas from Pierre d'Ailly, who drew his largely from Roger Bacon, who, in turn, was deeply indebted to the early Fathers.

The historical characteristic of medieval studies ceases to be extraordinary when we recollect that, during the Dark Ages at any rate, men of learning were so deeply concerned to keep their cultural legacy intact that they had little mind

[1] *op. cit.:* IX. Cf. Macrobius: *Commentarius*, II, 5.

[2] *Vide Examen critique de l'histoire de la géographie du nouveau continent . . .*, Vol. II.

or opportunity to enrich it. 'Europe, threatened by perils from every margin, was not safe. The most urgent duty of society was to preserve. The key to much of medieval scholarship was the text "Gather up the fragments that none may be left".'[1] Those who felt competent to impart any kind of instruction accordingly gave themselves to the preservation of such fragments of knowledge as they had inherited from the Ancients. Maintenance, rather than expansion, was the watchword. The prevailing mood was consequently one of acquiescence in authority. Past opinions came to assume greater importance than present realities, and the study of them was the only accredited way to arrive at the truth. To reject them was to reject light. Nor was this a very unnatural attitude of mind, for, as Whewell has remarked

> deference to the authority of thoughtful and sagacious men, a disposition which men in general neither reject nor think they ought to reject in practical matters, naturally clings to them, even in speculation. It is a satisfaction to us to suppose that there are, or have been, minds of transcendent powers, of wide and wise views, superior to the common errors and blindness of our nature. The pleasure of admiration and the repose of confidence are inducements to such a belief.[2]

Reasons such as these help us to see why in the early centuries of the Middle Ages the work of observation was in a large measure superseded by collection and analysis; why experimenters were replaced by commentators; why criticism took the place of induction and why, instead of geographical exploration, we had encyclopaedic scholarship. What is extraordinary is that so much should have been salvaged from the wreckage of the ancient world and that it should have been employed so wisely.

[1] H. C. Darby in *Scottish Geographical Mag.*, 1933, Vol. XLIX, p. 323.
[2] *History of the Inductive Sciences:* Vol. I, p. 204.

CHAPTER III

THE RISE AND FALL OF MOSLEM GEOGRAPHY

IF any one, in the opening years of the seventh century, had had the audacity to prophesy that within a decade or so some unheralded, unforeseen power, from the hitherto barbarous and little known land of Arabia, would arise and hurl itself against the only two world-powers of the age, falling heir to the one, the Persian, and robbing the other, the Byzantine, of its fairest provinces, he would undoubtedly have been a laughing-stock. Yet this was exactly what happened. No sooner had the Prophet died (632) than Arabia was transformed, as if by magic, into a nursery of heroes the like of whom would be hard to match. The military campaigns of Al-Walīd and Al-'Ās which ensued in Iraq, Persia, Syria and Egypt are among the most brilliantly executed in the history of warfare and bear favourable comparison with those of Hannibal, Alexander or Napoleon.[1]

While any rehearsal of their exploits would be out of place in this chapter, it is difficult to view the geographical achievement of the Arabs [2] in their proper perspective without reminding ourselves of the spectacular nature of this early advance.

The dates tell their own tale.

635 Damascus falls.
637 Jerusalem capitulates.

[1] *Vide* P. K. Hitti: *History of the Arabs*, p. 142.

[2] No matter what his nationality may originally have been, the follower of Mohammed was invariably called an Arab during these centuries. It is in this 'religious' sense that the word is employed in the following pages.

638 Mesopotamia is reduced.
c. 639 The Islamic era is established.[1]
640 Egypt is conquered.
643 Tripoli submits.
698 Carthage is subjugated.
711 The conquest of Spain is begun.

From a cultural standpoint the conquest of Persia and Syria was particularly significant: for in these countries the Arabian Moslem, who brought with him from the desert a keen sense of intellectual curiosity, a voracious appetite for learning and many latent faculties, found himself heir to older and superior cultures. These he was not slow to adopt. Within a century of the establishment of Baghdad he was in possession of the chief philosophical works of Aristotle, the leading Neo-Platonic commentaries, and most of the medical writings of Galen, as well as Persian and Indian scientific works.

For the time being, however, this unique inheritance profited him little, save only in as far as it enabled him to determine astronomical positions more precisely and to be more accurate in his horoscopic calculations. The explanation of his failure to capitalize these newly acquired gains is probably twofold. Firstly, his initial ignorance of the Greek language. (The translators employed by the eighth-century Caliphs of Baghdad were Jews and Nestorians.[2]) Secondly, and more important, his subservience to obscurantist religious views which prevented him from viewing the steadily growing wealth of information in its real light. Just as the early Fathers of the Christian Church laid down the tenets of orthodox belief on matters material as well as spiritual, with the Bible as their text-book, so the doctors of the Koran

[1] 'Nothing speaks more for the explosive growth of Islam than the successful establishment of this era only seventeen years after the initial event. It took from five to ten centuries to establish the Christian era.' —G. Sarton: *Introduction to the History of Science*, Vol. I, p. 464.

[2] These Nestorians, Syrians for the most part, translated first from Greek into Syriac and then from Syriac into Arabic.

legislated on the same matters with a similar vehemence. Among the sayings of the Koran which helped to fashion early geographical thought is the twice-recurring statement that God has separated 'the two seas' by an insurmountable barrier. These words were interpreted as an allusion to the Mediterranean Sea and Indian Ocean (including the Red Sea). And on the authority of contemporaries of the Prophet, many curious opinions were held concerning the length of the world, its parts, the sources of the Nile and similar topics. Among them we meet with the comparison of the world to a bird, whose head is China and whose tail is North Africa ![1] Mainly for reasons such as these there is little to choose between Arab and Christian scholarship for the first 150 years or so of the new era. But in justice to the Arab scholars, it is only fair to say that their prejudices were seldom as blind as those of their Christian contemporaries.

Secular science found its first entrance among the Arabs in the days of the early Caliphs of the Abbasid dynasty, whose ways and thoughts had been moulded by their residence in Persia amid the influences of an older creed, and of ideas which had, in the last resort, sprung from the Greeks. In the eighth century (*c.* 766) the seat of empire was transferred to Baghdad, on the highway of oriental commerce, and the distant Khorasan became the favourite province of the Caliph. Then was inaugurated the period of Persian supremacy, during which Islam was laid open to the full current of an alien culture. The incitement to secular study came, however, not from the people but from the prince. It was in the light of court favour that the colleges of Baghdad and Nīshāpūr first attracted students from the valleys of Andalusia and from the uplands of Trans-Oxiana. Al-Mansūr, the second of the Abbasids, encouraged the appropriation of Greek science: but it was Al-Ma'mūn[2] who deserves in the Mohammedan Empire the same position of royal patron

[1] *Vide* J. H. Kramers in *The Legacy of Islam* (edited by Sir Thomas Arnold), p. 83.

[2] *Fl. c.* 813–33.

and benefactor as is held by Charlemagne in the history of the Latin schools.

Early in the ninth century (*c.* 830), Al-Ma'mūn founded at Baghdad a scientific academy, called 'House of Wisdom', a library and an observatory. He began, at the same time, to collect Greek and Syriac manuscripts, among those procured being a chart fixing the latitude and longitude of every known place. He encouraged scholars of all kinds and creeds. Jews and Nestorians with linguistic ability were held in high esteem and made translations of many of the Greek scientific writings. 'The barbarism of the first believers was passing away, and Mohammed's words were recalled: "Seek knowledge, even in China".'[1]

Among the earliest works to be translated into Arabic from Greek were the major writings of Galen and Hippocrates and Ptolemy's *Quadripartitum*, by Al-Batrīq (*c.* A.D. 800). The *Elements* of Euclid and the *Almagest*, the great astronomical work of Ptolemy, may also have been translated about the same time if a report in Al-Mas'ūdī is correct.[2] But many of these early translations appear to have left much to be desired, for they were revised, in some cases remade, within a decade or so. Of all the 'classics', the works of Aristotle and Ptolemy were singled out for special honour and rapidly became the pattern of Arab vernacular writings. Of the two men, Aristotle enjoyed the greater popularity—a fact due, partly, to the importance which Avicenna (= Ibn Sīnā, *c.* 1000) and Averroes (= Ibn Rushd, *c.* A.D. 1150) attached to his works in their commentaries on them, and partly to their popularization in a tenth-century compilation by members of a religious fraternity. From these men, the Brothers of Sincerity or the Faithful Brethren of Basra, as they were called, there has come down to us an encyclopaedia of the sciences of that day. It consists of fifty-one treatises, the contents of which are of such varied nature and origin that the compilers have not succeeded in establishing a com-

[1] C. R. Beazley: *Prince Henry the Navigator*, p. 18.
[2] *Vide* P. K. Hitti: *History of the Arabs*, p. 311.

plete harmony among them. It was the first (and most notable) attempt among the Arabs to effect a universal reconciliation between knowledge and faith, and although it failed, as was inevitable, it exercised an important influence on the Islamic world—an influence to which eloquent testimony is borne by the existence of numerous manuscripts of the work. Thus it came about that to the Arabs of the eleventh and twelfth centuries, Aristotle represented and summed up Greek philosophy, even as Galen became to them the code of Greek medicine. From first to last Arabian philosophers made little or no claim to originality. Their aim was to propagate the truth of Aristotelianism as it had been delivered to them. It was with them that the apotheosis of Aristotle began, and from them the belief that in him human intelligence had reached its limit passed to the later schoolmen.[1] But the *Almagest* and *Geography* of Ptolemy also gained for themselves enviable positions of authority, being translated and copied with increasing frequency from the ninth century onwards. In addition, the *Almagest* provided the basis of such semi-original works as Alfragan's (= Al-Farghānī) *On the Elements of Astronomy*, Al-Battānī's *On the Movements of the Stars*, and Ibn Yūnus' *Hakimite Tables*; while the *Geography* was used as the model for *The Book of the Description of the Earth* (or *The Face of the Earth*) by Al-Khwārizmī and a similar work—with the same title—by Al-Battānī.

But these models were by no means followed as slavishly as those fulfilling a similar function in Christendom. With the vast extension of their empire, the *Weltanschauung* of the Arabs was necessarily widened. Their knowledge was continually being advanced by means of journeys undertaken to collect traditions, or for purposes of administration and trade, or simply to satisfy curiosity, to say nothing of pilgrim travel. Consequently new subjects of inquiry came up, together with new modes of treatment. It is therefore not surprising to find Arab scholars, with all their consideration

[1] *Vide Brit. Enc.*, 11th edition, Article *Arabian Philosophy*.

for the feelings and traditions of the past, submitting their authorities to severe, critical analysis and revising them where expedient. Al-Khwārizmī, for instance, substantially improves Ptolemy's *Geography* in his *Face of the Earth*, as regards both the text and the maps,[1] while several writers after his time give themselves to the revision of the Ptolemaic system of astronomy with which there was a growing dissatisfaction. This movement, initiated by Ibn Bajja, was considerably elaborated by Jābir Ibn Aflāh, the greatest astronomer of the twelfth century. Ibn Aflāh wrote a treatise entitled *The Correction of the Almagest* which exerted a deep influence upon medieval thought through its Latin and Hebrew translations,[2] and Alfragan altered Ptolemy's figure for the length of a geographical degree.[3] This anti-Ptolemaic tendency was, however, reactionary rather than progressive. The commentators had really no idea of the fundamental defects of Ptolemaism, whether geographical or astronomical, and their criticisms were largely superficial. Still they were valuable for all that. It was the attitude of these writers to authority rather than the validity of their arguments, or the intrinsic superiority of their findings, that was most significant for the future of scientific inquiry.

Alongside these theoretical studies, which contributed little or nothing to the geographical knowledge of the contemporary world, a literature of a more topographical and descriptive character was steadily accumulating. The underlying cause

[1] This work was accompanied by a map of the world executed by Al-Khwārizmī and sixty-nine other scholars at the instigation of Al-Ma'mūn (the first of its kind to be drawn in Islam). Mas'ūdī (*vide infra*) consulted this map.

[2] Gerard of Cremona was responsible for the first Latin version of it, *c.* 1187.

[3] Alfragan's figure was $56\frac{2}{3}$ miles, and Ptolemy's $62\frac{1}{2}$. However, owing to the fact that Alfragan's mile was expressed originally in terms of cubits which were of two kinds, great and small (18″ and 15″ respectively), two figures for the circumference of the earth were commonly attributed to him, namely, 24,480 miles and 20,400. The former of these was considerably nearer the truth than Ptolemy's figure of 22,500 miles.

of this development—without parallel in contemporary Christendom—is not difficult to find. To begin with, pilgrimage for the Mohammedan was obligatory. Unless he were incapacitated by illness or circumstance, he had to make the journey to Mecca before he died. Accordingly there flowed into Arabia a continual stream of visitors from every corner of Islam. Some of them came from Central Asia and beyond, others hailed from the western end of the world. To all of them were offered unrivalled opportunities for the friendly exchange of opinions not only in the sacred place itself but in other Moslem cities as well as along the pilgrim way. Few travellers, needless to say, were so narrow-minded that they thought only of their main object. 'The pilgrimage was the accomplishment of a religious duty, but it was also a unique vacation, an initiation into the great world of Islam, and for scholars, the exact equivalent of our journeys for study abroad.'[1] For this reason it became a powerful factor not only in promoting religious unity, but also in strengthening the ties of commerce between all Mohammedan countries and in disseminating a knowledge of all parts of the world. The itineraries and topographical description of the ninth and following centuries were therefore a natural by-product of the 'hajj', as the pilgrimage was called. Broadly speaking, these works contain accurate and interesting eye-witness information about the countries visited. One of the earliest of them was the *Book of Roads and Provinces* of Ibn Khurdādhbih (*fl. c.* 850). The author, an official in the Central postal service at Samarra near Baghdad, was commissioned to undertake its compilation by the Caliph Al-Mo'Tamid. In addition to describing the trading stations, post-relays and revenues of each province in Islam, it contains a geographical preface of a more general character. Quoting from an earlier Arabic source, Ibn Khurdādhbih assures us in this preface that

> the earth is round like a ball and placed in the centre of the heavens, just as the yolk is in the middle of an egg. . . . [It] is 9,000

[1] G. Sarton: *op. cit.*, Vol. II, part I, p. 35.

parasangs in circumference.[1] . . . The northern quarter of the globe is inhabited, while the southern is desert, because of the great heat that prevails there. The other half of the earth, opposite to ours, is uninhabited. The northern and southern quarters are divided into seven climates. Ptolemy says in his book that in his time there were 4,200 towns in the world.[2]

Similar works followed: the *Book of Countries* of Al-Ya'qūbi (*c.* 891), which struck a new note in emphasizing topographical and economic detail; the *Figures of the Climates* of Al-Balkhī (921), consisting chiefly of regional maps; the *Routes of the Provinces* of Al-Istakhrī (*fl.* 950), really a more elaborately illustrated edition of the former. This was further revised and enlarged at Al-Istakhrī's request by Ibn Haukal (*c.* 977) under the name of *Book of Roads and Provinces.* To the same category belongs the more original work of Al-Maqdisī (Al-Muqaddasī) entitled *The Best of Classification for the Knowledge of Climate* (*c.* 985). In the following century, Al-Bakrī, the earliest of the western Moslem geographers whose works have survived, continued the sequence by writing *The Book of Roads and Kingdoms* (*c.* 1050).

The system initiated by Ibn Khurdādhbih and Al-Balkhī paid little attention to countries outside Islam and made the text largely a description of the accompanying maps. A perusal of these works makes it clear that the authors took great pains to obtain in each locality accurate information concerning the climate, the products, the state of trade, the coinage, weights and measures and the general characteristics of the inhabitants. Ibn Haukal puts his aim thus:

to describe the various climates and regions of the face of the earth, comprised within the circle of Islam, or Mohammedanism, and their several divisions, in such a manner that every remarkable place belonging to each region shall be noticed, and all the boundaries and territories depending on them, their districts, cities,

[1] *parasang* = 12,000 cubits = 3 miles, 1 cubit equalling 18 inches.

[2] *Vide* M. J. de Goeje's translation, Vol. VI, pp. 2–3 of *Bibliotheca Geographorum Arabicorum.*

mountains, rivers, lakes, and deserts . . . and to delineate [them] on maps.[1]

That he achieved his aim may be surmised from an extract taken almost at random—his description of the Caspian Sea region.

The Western side of this sea belongs to Deilman and Taberistan and Gurkan, and its borders; and part of it is bordered by the deserts of Khuarezm. On the Eastern side is Aran and Moukan and the territories of Serir and part of the deserts of Azziah: and on the north it has the desert of Azziah, to the territories of Siah Kouh: and on the south, Bakeil and Deilman and the neighbouring places. This sea is not connected with any other; and if a person wishes to make a tour completely round it, nothing will impede him but a few rivers which fall into it from various quarters. The waters of this sea are bitter and dark-coloured: its bottom is a blackish sea, differing in this respect from the Sea of Kolzum [2] or of Oman or of Pars.[3] This sea of Pars is of such clear water that one may see the white stones at the bottom; but the waters of this Sea of Khozr [4] are dark-coloured, and in it there are not found any such things as pearls, or coral or similar marine productions. It is, however, much frequented by the ships of merchants who traffic from one town to another: and it affords much fishing. In this ocean there are not any inhabited islands as in the Sea of Pars or of Roum [5]: but there are many trees and forests. One island is considerable, with a spring of water, and many trees: and there is another large island on the borders of Lekzan, which also has fresh water. To this island they bring cattle from Berdaa in boats, and turn them out to graze and leave them until they become fat.[6]

Al-Muqaddasī is even more ambitious. Not content with recording the results of his own observations, he also puts down what he has heard from trustworthy people and what he has read in books. He attaches, however, a greater value to the

[1] *Vide* W. Ouseley's Translation, pp. 1–2.
[2] i.e. Red Sea. [3] Persian Gulf.
[4] i.e. Caspian Sea. [5] i.e. Mediterranean Sea.
[6] W. Ouseley's Translation, pp. 183–5.

first than to the rest, for he maintains that the only real knowledge is that gained by research and inquiry, not by faith in tradition or by deductions from the intellect. Accordingly, such things as the faulty geographical statements of the Koran he explains by the limited physical horizons of the early Arabs, to which Allah must have seen fit to adapt himself.

From itineraries to text-books, utilizing both the old academic studies and the new eye-witness narratives, was no far cry for the Arab scholar-pilgrim; at least, not for a man of the calibre of Al-Masʿūdī, for here is a man of the tenth century with a fifteenth-century renaissance mind. He is interested in and appreciates everything that concerns humanity. Everywhere he is learning something from the things he sees and the people he meets; he records information about events and views of men without fear or prejudice. As a result of his travels—travels stretching from Galicia to China and from Madagascar to the heart of Russia—Masʿūdī finds himself compelled to question not only many of the accepted facts of his day, but the theories also, particularly Ptolemy's. He asks, for instance: Is Africa surrounded by sea? What is the shape of South-East Asia? Is there an unknown Southern Continent? With commendable courage he discards the Ptolemaic concept of an Africa linked by land with South-East Asia, and regards it as sea-girt and separated by a narrow strait from a southern landmass—the 'terra Australis nondum cognita' of later days. All the seas, he avers, 'are connected and uninterrupted. The first is the Abyssinian Sea,[1] then the Mediterranean Sea, the Pontus,[2] the Mayotis,[3] the Khazarim,[4] and the Ocean which is also called the Green Sea, the surrounding sea and the Dark Sea.'[5]

On the purely descriptive side, Masʿūdī's work is remarkable

[1] i.e. Indian Ocean. [2] i.e. Black Sea.
[3] i.e. Sea of Azov. [4] i.e. Caspian Sea.
[5] *Meadows of Gold* . . . (Translation by A. Sprenger), p. 297. In the next sentence or so, however, Masʿūdī qualifies this statement by saying that 'they are in no connexion whatever with the Caspian Sea'. Cf. Ibn Haukal's description, *ante* p. 52.

for some rather paradoxical things. On the one hand, there is his catholicity—he neglects no source of information and describes Moslem and non-Moslem countries with impartiality; and his scientific curiosity—he describes the earthquake of 955, the waters of the Dead Sea, and the origin of seas in general with the detachment of a present-day observer. On the other hand, he repeats, without venturing to challenge the rightness of it, the following palpably erroneous estimate of the earth's size:

Men who are possessed of an extensive knowledge of our globe say that the Abyssinians and [people of] Sudan occupy a country which is seven years' journey, that Egypt forms the sixtieth part of the country of the Sudan and that the country of the Sudan forms only a small portion of the surface of the globe, the dimensions of which are equal to five hundred years' journey. One-third of it is inhabited, one-third is desert and the remaining third is sea-covered.[1]

A similar fusion of ancient tradition and modern knowledge is attempted by several other Arab geographers of the time. Here, for instance, is Al-Bīrūnī's[2] discussion of the limits of contemporary knowledge.

The Western Sea which we call Ocean has been named the Environing Sea, because in effect its shores commence at the extremity of the countries which it touches on the southern side, opposite the Negro lands and runs past successively the shores of Audagost, Sous-Alacsa, Tanger and Tahurt advancing towards the shores of Andalusia, Galicia and the countries of the Slavs. Then, in turning north, it goes round the inhabited world and countries behind unknown mountains and lands desolate because of the cold and flows east by an unsuspected route as far as the Oriental Sea which marks the limit of the inhabited world. It has not been

[1] *Meadows of Gold* . . ., p. 377. And this, in spite of the fact that Mas'ūdī reduces his Atlantic Ocean to extremely modest proportions. His Fortunate Isles (i.e. Canaries) abut on India! On the other hand, his Sea of Habasch (i.e. Indian Ocean) contains most of the water surface of the world.

[2] *Fl.* eleventh century.

described with the same exactitude as the Ocean.[1] We know, however, in a general way, that it continues towards the south just as the Ocean continues towards the north. The Southern Sea commences at China and flows along the shores of India towards the country of the Zendj.[2] The coasts to the north are inhabited but those which border it on the south remain unknown, for they have not been visited by any navigator and the people who live in the adjacent isles have not furnished us with any knowledge on this point. The extreme part which has been visited by the people in the Southern Sea on its west side is Sofala, in the country of Zendj. Navigators have not passed this limit, the reason being that the sea on the north-east penetrates into the land . . . while on the south-west, as if by way of compensation, the continent projects into the sea. This region is inhabited by the Negroes of the West and it stretches beyond the Equator to the mountains of Comr where the Nile has its source. Beyond this point, the sea penetrates between the mountains and valleys which alternate with one another. The water is continually set in motion by the ebb and flow of the tide, the waves for ever surging to and fro, so that ships are broken in pieces. This is why the sea is not navigated. But this does not prevent the Southern Sea from communicating with the Ocean through a gap in the mountains along the south coast [of Africa]. One has certain proofs of this communication although no one has been able to confirm it by sight. It is because of this inter-communication that the habitable part of the world has been placed in the centre of a vast area environed on all sides by the sea.[3]

The general meaning of this passage being clear, the following deductions may readily be drawn:

1. That, by the eleventh century, the Arabs were in possession of information about the east coast of Africa to at least 20° S., i.e. as far as Sofala.

2. That statements concerning regions to the south are entirely conjectural. At the same time, the mention of ships being broken in pieces may imply a second-hand acquaintance with the Mozambique channel.

[1] i.e. Atlantic. [2] i.e. Zanzibar region.

[3] *Vide* J. T. Reinaud: *La Géographie d'Aboulfeda*, Vol. II, pp. 14–15.

3. That the peninsular nature of the African continent was maintained despite opposition from the followers of Ptolemy and the lack of confirmatory evidence.

4. That the existence of a southern antipodal landmass was unquestioned.

5. That very little was known either of Northern Europe or of Eastern Asia by the Arabs of this period.

Elsewhere in his writings Al-Bīrūnī discusses intelligently such wide-ranging topics as the earth's rotation on its axis, the working of natural springs by the hydrostatic principle, and the origin of the Indus valley which he rightly surmises to have been an ancient sea basin filled up with alluvium.

The most notable example of the fusion of ancient and modern geography is found in the work of Al-Idrīsī, a man whom genius and circumstance served right royally. Born in Ceuta in 1099, this great geographer travelled widely in Europe, Africa and the Levant, before being persuaded by Roger II, the Norman King of Sicily, himself a man of some learning, to settle at the court of Palermo. While there, he undertook to write a geographical treatise entitled, rather fulsomely, '*Amusement for him who desires to travel round the world.*' About the same time (*c.* 1154) he also made a celestial sphere and a representation of the known world in the form of a disk, whereon was inscribed 'all the circuit of the known world and all the rivers thereof'. He was obviously in a very advantageous position for carrying out all this work, as Sicily was the rendezvous of navigators from Mediterranean, Atlantic and Northern waters, and so was in closer touch with contemporary life than most of the Islamic centres of culture. A typical instance of his knowledge of current events is supplied by the account of the Lisbon 'wanderers'.[1] This, in addition to its intrinsic value as being the earliest recorded voyage after the rise of Islam, deliberately undertaken for the purpose of exploring the Western Ocean, provides an interesting sidelight upon the breadth of the writer's outlook. But in coming by

[1] *Vide* C. R. Beazley: *The Dawn of Modern Geography*, Vol. III, pp. 532–3 (note).

such knowledge he owed a very great deal to the generosity of his patron, who employed men in almost every part of the civilized world to glean materials for him.

The plan of this work is simple, though somewhat artificial. After a brief description of the earth as a globe, which he computed to be 22,900 miles [1] in circumference and judged to remain 'stable in space like the yolk in an egg', and of the hemispheres, climates, seas and gulfs, he launches out on a lengthy account of the regions of the earth's surface. Taking the seven main Ptolemaic *climata* in order, he divides each of them into ten longitudinal sections, a fashion started by earlier Islamic astronomers. These seventy sections he describes in detail, illustrating each description with a map. When put together these maps constitute a rectangular world map much after the Ptolemaic pattern.[2] Notwithstanding the arbitrariness of the divisions, his description of them is of exceptional merit when contrasted with the Christian, or for that matter, with the other Moslem literature of the period. This is due partly to the wealth of detail he furnishes (especially in connexion with Africa), but mainly to his scientific method, the co-operation of numerous observers, and his critical correlation of their observations.

We will examine his knowledge of Africa by way of example.

The first division of the first climate commences to the west of the Western Sea which is called the 'Sea of Darkness'. 'In this sea are two islands named Al-Khālibāt[3] where Ptolemy began to count longitudes and latitudes (*sic*) . . . nobody knows of habitable land beyond that.'[4] In this southernmost section he places a number of important towns including the problematical Oulil[5] which, he tells us, 'is situated in the sea

[1] Cf. 22,500 miles of Ptolemy's estimate.

[2] *Vide infra*, pp. 66–7 and Plate No. V.

[3] i.e. Fortunate Isles.

[4] *Description de l'Afrique et de l'Espagne* . . . (Translation by R. Dozy and M. J. de Goeje, p. 1 *et seq.*).

[5] Cape Timiris? *Vide* E. G. R. Taylor: 'Pactolus—River of Gold,' *Scottish Geographical Mag.*, Vol. XLIV, p. 137.

not far from the shore' and is renowned for salt. Much of the trade in this commodity with the Sudan was done with the help of ships which carried it from the town of Oulil a day's journey

> to the mouth of the Nile [1] and mounted the river as far as Silla, Tacrour, Barisa, Ghana . . . [and] to all the Sudanese towns. The greater part of the country is only habitable on the borders of the Nile for the rest of the country . . . is desert and uninhabited. There are arid wastes where one must walk two, four, five or twelve days before finding water. . . . The people of Barisa, Silla, Tacrour and Ghana make excursions into Lamlam [2] bringing the natives into captivity, transporting them to their own country and selling them to the merchants.

In the second section of his first climate, he describes, among others, the lost city of Ghana, farther to the east, as

> the most considerable, the most densely peopled and the largest trading centre of the Negro countries. . . . From the town of Ghana, the borders of Wangara are eight days' journey. This country is renowned for the quantity and the abundance of the gold it produces. It forms an island 300 miles long by 150 miles wide : this is surrounded by the Nile on all sides and at all seasons. . . . The greater part [of the gold] is bought by the people of Wargalan [3] and by those of Western Maghrib.[4]

Following the Nile still eastwards, 'we find the nomadic Berbers who pasture their flock on the borders of a river flowing from the east, debouching into the Nile stream'. Beyond, in the fourth section of his first climate, we come to

> the place where the two Niles separate, that is to say, first, the Nile of Egypt which crosses the country from the north to south, and second, the branch which flows from the east towards the

[1] i.e. Senegal R., or Nile of the Negroes.

[2] Lamlam (or Lemlem) is probably to be identified with the hinterland of the Ivory and Liberian coasts. *Vide* E. G. R. Taylor : *loc. cit.*, pp. 139 *et seq.*

[3] i.e. Wargla.

[4] i.e. Morocco.

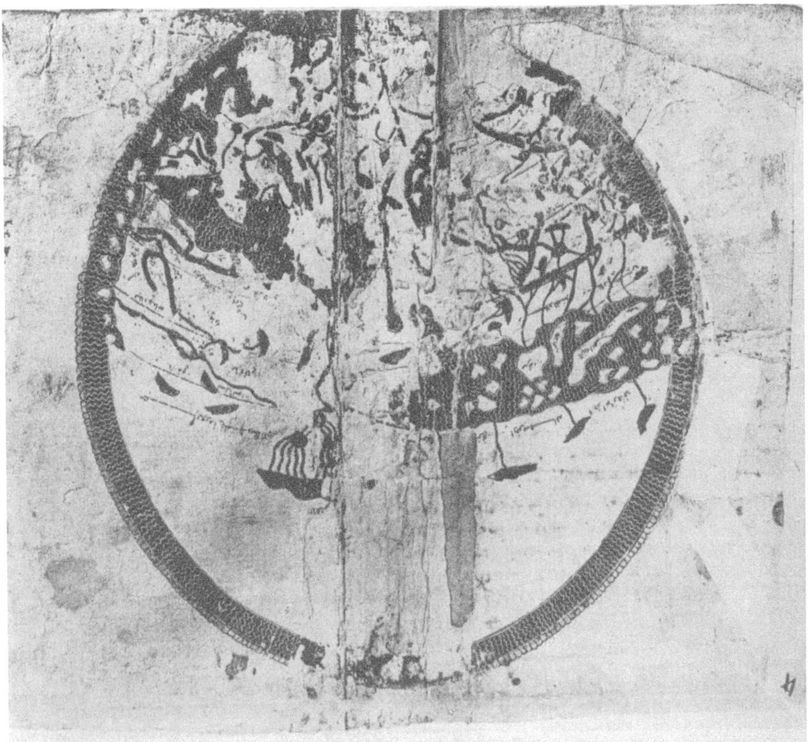

V

WORLD MAP ATTRIBUTED TO AL-IDRĪSĪ—FROM TWELFTH (?) CENTURY MS.

(*Nat. Bibl. Paris, No.* 2,221—*Arabic*)

western extremity of the continent. It is on this branch of the Nile that most of the large towns of the Sudan are situated.[1]

These few extracts are characteristic of Idrīsī's method and his matter. From them we see, for instance, that Ptolemy's authority no longer commands unreserved approval; Ptolemy places the source of the Nile south of the Equator, in the Mountains of the Moon, and has no sympathy with the idea of a dual Nile. We see, further, that there was already, by the twelfth century, a regular commercial exchange between the Mediterranean Sea and the Sudan, and that authentic information concerning these southern lands was beginning to filter through to the European centres of learning. When we recall that the *exact* hydrography of the land of the 'Western Nile' was not discovered until the nineteenth century, Idrīsī's narrative assumes a profound importance. The authenticity of many of the places he mentions is indisputable. Thus Ghana (situated near Timbuktu), Silla (possibly Ysilgam of the Valseccha Portolan chart of 1434[2] and Tacrour (Tekrour on the Senegal) were, for a time, flourishing centres of Moslem culture. The reference to Wangara implies a knowledge of the flood region of the Niger above Timbuktu,[3] and the mention of the salt trade of Oulil suggests that there were, in Europe, faint glimmerings of knowledge about the Senegalese coast, even as early as the mid-twelfth century.

In view of its 'modernity' and high intrinsic worth, it is difficult to understand why Idrīsī's work, composed as it was at the chronological and geographical point of contact between the Islamic and Christian civilizations, remained so long unutilized by Christian scholars in Sicily, Italy or other Christian countries, until we remember that the primary—we might almost say the sole—interest of the Latin West in Arabic literature centred in the preparation of calendars, star tables

[1] *Description de l'Afrique* . . ., p. 17.

[2] For a reproduction, *Vide* Charles de la Roncière *La découverte de l'Afrique au moyen âge*, Vol I, Plate No. XII. Some authorities give the date as 1439.

[3] *Vide* E. G. R. Taylor: *loc. cit.*, p. 141.

and horoscopes, and the recovery of ancient lore. It was not much concerned in the twelfth century with the descriptive geography of Africa or Asia.[1]

After the time of Idrīsī, Arab geographical literature—excluding purely travel literature[2]—cannot claim any great originality. Ibn Sa'īd (1214–74) spends his time recording other people's travels; Al-Khazwīnī (*fl. c.* 1250) writes a natural history in the Plinian tradition; Ibn Yākūt (*fl. c.* 1225) compiles a *Geographical Dictionary* based largely on the writings of earlier geographers, and Abū'l Fidā (*fl. c.* 1300–25) produces an encyclopaedia. This last work is typical of the geographical thought of the period and provides good evidence of the persistence of the higher Moslem culture amidst intellectual depression. As a Sultan (of Hamat) it is to be imagined that Abū'l Fidā was in fairly close touch with contemporary happenings, and so was able to gather at first hand his material for a geographical treatise. But this does not appear to have been the case, for we find him drawing largely upon earlier authorities. Of those he mentions by name—no less than thirty-six—Al-Bīrūnī is easily the most revered. Yet Abū'l Fidā did not hesitate upon occasion to revise and amplify the theories of his authorities. Thus he was disinclined to follow Al-Bīrūnī in his theory of the habitability of the Southern Hemisphere. 'The Equator and the Great Circle passing through the two Poles divide the earth into four parts. It is one of these quarters to the north which forms the habitable world, for the other three-quarters are unknown to us and it is widely held that they are covered with water.'[3] He verifies this statement, adducing as a proof the fact that 'only in a very small area do the shadows at noon, with the sun overhead at the Equator, fall to the South,' e.g. in the country of Zanzibar.

[1] The first translation known of his work was published in Rome only in 1619, and then in a very much shortened form. The translator did not even know the author's name!

[2] *Vide infra*, chap. v.

[3] J. T. Reinaud: *La Géographie d'Aboulfeda*, Vol. II, p. 5.

The sea he regards as surrounding the earth on all sides. ' We know the North-West and South-East coasts, but nobody has ever visited the South-West, and we still have no exact idea of the lands to the North-East.' 'The Southern quarter is not inhabited by reason of the proximity of the sun', but Abū'l Fidā admits that this view is not shared by all the astronomers.

From his description of the Environing Sea, we gain some insight into his conception of the southern limits of the *oikoumenē*.

The west coast of the Environing Sea . . . carries the name of Ocean. Here are found the Eternal Isles [1] at a distance of ten degrees from the continent . . . [the sea] stretches from the coasts of Magreb-Alecsa [2] towards the South as far as the desert of Lamtounas. This desert which is occupied by the Berbers extends between the limit of Magreb and that of the country of the Negroes.[3] Then the sea continues South along uninhabited and unfrequented shores, as far as the Equator. Then it turns towards the East behind the Comr Mountains [4] where the Nile takes its source, and washes the southern coasts of the world. Thence it continues in the direction of the East past uninhabited countries, behind the country of Zendj.[5] There it takes a North-East direction and stretches to the sea of China and India. . . . According to the Sheriff Idrīsī, the water of the Environing Sea, on the South coast, is viscous and this, he says, is because the sun . . . evaporates the volatile atoms in the water.[6]

The above quotation—and it is fairly typical of Arab geography in the later Middle Ages—shows that traditional ideas were still crippling the growth of a native geographical science. The same passage shows, equally clearly, that Abū'l Fidā's ideas on the southern extension of Africa are a retrogression upon those of Al-Bīrūnī already set out. The latter advances the not unreasonable postulate of inter-communica-

[1] i.e. Canary Islands. [2] i.e. Morocco. [3] i.e. The Sudan.

[4] Fouta Djallon Mountains of West Africa ? The River Niger (i.e. the Western Nile') has its source in them.

[5] i.e. Zanzibar region.

[6] J. T. Reinaud : *op. cit.*, Vol. II, pp. 23–4.

tion between the 'Abyssinian Sea' and the Ocean (i.e. Atlantic) in the region of Madagascar. For this Abū'l Fidā substitutes the resuscitated and partially modified [1] Ptolemaic theory—a theory completely at variance with the same scholar's opinion expressed elsewhere. For in the section on Bilad Al Soudan (the Country of the Blacks) he says 'the town of Daghouta is the last in the country of Sofala and the most remote of an inhabited part of the continent—109° long: 12° S.' [2] The ascription of habitable land to Africa south of the Equator is, of course, not readily squared with the statement that the Environing Sea 'turns towards the East' along the latitude of the Equator.[3]

Such was the confused state of Arab geographical thought in the early fourteenth century. During the next hundred years or so there were growing symptoms of dissatisfaction with the accredited theories and an increasing reluctance to accept classical authority. Ultimately this brought about the overthrow of the Ptolemaic tradition. By a strange and unfortunate coincidence, at the epoch when this courageous move was being made, Western scholars were ceasing to hold Arab scholarship in as high regard as hitherto. For Ptolemy's *Geography* and *Almagest* had been rendered accessible to them in Latin, and these treatises, so far superior to anything they yet possessed, won their allegiance and thereby helped to retard the growth of accurate geographical concepts.

Until then (i.e. *c.* 1410), Christendom had accepted the Moslem views with deference, though not without a certain amount of disinclination, born of religious antipathy. Henceforth she was determined to cull her knowledge from the Ancients at first-hand.

The original contributions of Arabic geographers to their science during these centuries were confined mainly to the

[1] By Idrīsī who conceived the Indian Ocean to be only partly enclosed. *Vide* Plate No. V.

[2] J. T. Reinaud: *op. cit.*, Vol. II, p. 208.

[3] Unless it may be interpreted to mean any direction east of south.

mathematical side, for, apart from an elaboration of the ancient theory of cosmic cycles,[1] they consisted of improvements in the technique of astronomical observations and cartography. The Arabs had early felt the need for accurate determination of position, if for no other reason than that the correct orientation of their mosques depended upon such knowledge. According to Mohammedan law, these had to face towards Mecca. Astrology, which was held in high esteem in Islam, also encouraged the same type of investigation, for, in order to cast a horoscope, it is necessary to know what stars are overhead at a given moment, and, to ascertain this, it is necessary to know the latitude and longitude. Accordingly, many a Moslem scientist gave himself to the improvement of methods of finding these geographical co-ordinates, and with commendable results, for the figures normally employed in their tables are considerably more exact than those given by Ptolemy. Al-Zarqālī, a Spanish Moslem of the twelfth century, gives the length of the Mediterranean Sea in his commentary on the *Toledo Tables*,[2] as 42°, which is almost exact,[3] while Al-Marrakūshī gives a list of 135 co-ordinates which, for accuracy, deserves to be regarded as one of the most important contributions of the Moslems to mathematical geography.

The methods employed by Al-Zarqālī are set out in the *Canones*, or Canons, as his commentary was called in its Latin form. They are twofold : in one case the sun is used, in the other a fixed star. The solar method is as follows : When the sun is overhead at the Equator, the observer must measure its elevation in degrees above the horizon at noon with the astrolabe. This figure subtracted from 90° will give him the latitude. If, however, he wants to find the latitude at some other time of the year than that of the equinoxes, he must ascertain the number of degrees separating the sun from

[1] *Vide* chap. vii, pp. 169–70.

[2] So named because astronomical and astrological tables had been compiled for that city shortly before the end of the eleventh century.

[3] cf. 62° of Ptolemy.

the celestial equator for that particular date. (This was easily found in the solar tables which were part of the astronomical tables on which Al-Zarqālī's Canons were a commentary). By adding this number to the number of degrees of the sun's elevation when the sun is south of the celestial equator, and by subtracting it when the sun is north of that equator, he will get the position of the sun as it would be at the equinox. From this the latitude can readily be found by subtracting from 90° as we have just seen.

The method of observing fixed stars is even simpler. All the observer has to do is to determine the elevation of the celestial pole above the horizon. To find this he selects some star which never sets and measures the elevations above the horizon of the two points where this star cuts the meridian at its upper and lower transits in its circular course around the celestial pole. The pole, obviously, is half-way between these two points and the elevation of the pole in degrees above the horizon is the same as the latitude.[1]

An ingenious, if somewhat more elaborate, method of calculating latitude was used by Al-Haitham,[2] who recommended using a bright fixed star for the precise determination of the altitude of the celestial pole, a star which in its upper culmination reached the zenith or at least came very close to the zenith. Then, very near the zenith, two so-called corresponding altitudes of the star to the East and West were taken with the astrolabe. In this way the influence of refraction, known to affect the measured altitudes of stars at greater distances from the zenith, could be eliminated. The time which elapsed between the moment the fixed star left its eastern altitude and the moment when it reached its western altitude was measured as accurately as possible with a water-clock. When the two observed heights, the interval between the observation, and (in case the star did not culminate in the zenith) the altitude of its culmination, had been measured, the

[1] For a fuller discussion *vide* J. K. Wright : 'Notes on the Knowledge of Latitudes and Longitudes in the Middle Ages', *Isis*, Vol. V, 1923.

[2] *Fl. c.* A.D. 1000.

latitude of the place of observation could be calculated.[1] It is not known how many calculations were made by such methods, but certainly some of the Arab findings are strikingly near the truth. The latitude of Baghdad was found by the sons of Mūsā Ibn Shākir to be 33° 20′ N.—a figure that is correct within a minute.

For the calculation of longitude the Arabs relied very largely upon eclipses of the moon. That the moon was used in the case of the *Toledo Tables* is certain, for Roger of Hereford, who adapted the Tables to the meridian of his own city, tells us that the time of the eclipse of 1178 was observed in Toledo, Hereford and Marseilles, and their longitudes in relation to Arin, the Arabic world centre, were in that way determined.[2] Only in one place is the so-called 'terrestrial' calculation of longitude mentioned. This is in the work of Al-Bīrūnī, who describes how, having determined accurately the shortest linear distance between two points, the latitudes of each and their mutual bearings, he proceeded to calculate the difference in longitude from the data thus acquired. The one result which Al-Bīrūnī claims to have arrived at by this method is remarkably good, for the figure given for the difference in longitude of Baghdad and Ghazna is 24° 20′.[3] The eclipse method is clearly set out for us in the *Hakimite Tables* of Ibn Yūnus, but the results were generally unreliable. Thus Hereford, London, Paris and Marseilles are given the following longitudes in relation to Toledo in twelfth- and thirteenth-century Tables.

[1] Al-Haitham's views did not, however, enjoy any great popularity in the Middle Ages. A disciple of Maimonides, the Jewish philosopher of the twelfth century, relates that he was in Baghdad on business when the library of a certain heretic was burned there. The preacher, who conducted the execution of the sentence, threw into the flames, with his own hands, one of Al-Haitham's astronomical works, pointing as he did so to a delineation therein given of the sphere of the earth—'an unhappy symbol of impious atheism.' *Vide* T. J. de Boer : *The History of Philosophy in Islam*, p. 153.

[2] Arundel MSS. No. 377, fol. 86 (British Museum).

[3] Actually 23° 50′. Ghazna 68° 17′, Baghdad 44° 27′.

Hereford	9° E,	giving a positive error of				7°	43′
London	4° E,	„	„	„	„		5′
Paris	12° E,	„	„	„	„	5°	40′
Marseilles	16° 30′ E,	„	„	„	„	7°	8′

While the Arabs were good astronomers they were poor cartographers. When they strayed from Ptolemy, as they frequently did, their maps became the vehicle for fancies rather than facts. Even the world-maps attributed to Idrīsī, carefully drawn as they were, and owing much to Ptolemy, betray a sad ignorance of the meaning of geographical co-ordinates as understood by the Alexandrine master.[1] Specimens of their map-makers' art, of both earlier and later date than his, are cast in an even cruder mould. They owe more to the late Latin school than to the Ptolemaic and are more akin to diagrams than maps. But in qualification of this indictment it is only right to say that their authorship is, almost without exception, disputable, and that there is no certain proof that any one of them is the work of the author in whose writing it occurs. Notwithstanding, it is now possible, thanks largely to the untiring researches of Konrad Miller [2] and Prince Youssouf Kamal,[3] to arrive at an estimate of Arabic cartography more satisfactory than any attempted, hitherto, by most western scholars. In general, their world-maps subscribe to the old classical notion of an ocean embracing the habitable earth which commonly assumes a circular form. Following Ptolemy, the habitable earth is divided into *climata*, or latitudinal zones, and extends only

[1] In his scholarly work, *Mappae Arabicae—Arabische Welt und Länder Karten des 9–13. Jahrhunderts*, Konrad Miller is of the opinion that the Ptolemaic influence has been very much exaggerated. Thus, for example, although Idrīsī adopted the *climata* system of regional treatment, in detail it has very little in common with Ptolemy's. Again, the whole delineation of the south coast of Asia and Malay differs widely from that of Ptolemy, most noticeable perhaps being the omission of any indication of the Malay peninsula, which is shown in exaggerated dimensions even by Ptolemy.

[2] *op. cit.*

[3] *Monumenta Cartographica Africae et Aegypti:* Tome III, 1926.

a little way into the Southern Hemisphere. Africa, thus foreshortened, is given an eastern prolongation, which, however, does not completely land-lock the Indian Ocean.[1] Communication is maintained between it and the Environing Sea by means of a narrow channel—as described by Al-Bīrūnī. Many maps reveal the Ptolemaic influence still further in that they show the sources of the Nile in equatorial latitudes,[2] but some others favour the Plinian idea of a western branch of the river rising in the Sahara. Other common characteristics are the disproportionate size of the inland seas and ocean gulfs, the frequent disagreement between the text and the drawing where the geographical location of towns, rivers and mountains are concerned, the habit of orientating maps with the North at the top and writing all the rubrics from the northern side, the absence of lines of latitude and longitude, the adoption of a datum point—under the appellation of Arin—through which passed the prime meridian,[3] and, occasionally, the placing of Mecca in an orbo-centric position. Towards the end of the fourteenth century the influence of the Portolan chart began to be felt and it was not long before the Arabs were drawing copies of the standard Italian and Catalan models.[4]

* * * * *

Brief as this survey has been, it can hardly have failed to convince us of the superiority of Arabic geography over the

[1] *Vide* Plate No. V.

[2] Miller (*op. cit.*) is again sceptical about the extent of Ptolemaic influence. In fact, he argues that a close scrutiny of Idrīsī's and Ptolemy's Nile system reveals more points of dissimilarity than of similarity and that Ptolemy's delineation is far more accurate.

[3] The western prime meridian was commonly supposed to be 90° from a mythical city, Arin (or Arim) by name, situated on the Equator midway between the eastern and western extremities of the known world. This city was said to have neither latitude nor longitude, and its meridian was arbitrarily established some ten degrees east of Baghdad.—*Vide* Al-Masʿūdī : *Meadows of Gold . . .*, *op. cit.*, p. 196, and J. K. Wright : *Geographical Lore of the Time of the Crusades*, p. 86.

[4] Youssouf Kamal : *Quelques éclaircissements épars sur mes Mon. Cart. Africae*, pp. 107–8.

contemporary geography of Christendom. The explanation of this contrast would seem not to be difficult. During the Middle Ages, as we have already observed, the Greek tradition of disinterested research was stifled in Western Europe by a theological dictatorship which bade fair, for a time, to destroy all hope of a genuine intellectual revival. Further, the Latin West had gradually drifted apart from the Greek East, thereby widening the already-present cultural cleavage. Meanwhile the Moslems were slowly unearthing the treasures of Greek and Persian wisdom, and in so doing they became fired with an enthusiasm to study them. Aided by their own native genius, by the keenest inter-regional competition—for Moslem culture radiated from a number of centres distributed all the way from Seville to Samarkand—and the stimulus of the classical models, they succeeded in advancing the cause of every known science before being overtaken by a tyrannical obscurantiscism.[1] About the same period, scholarship in Christendom was undergoing a similar fiery trial from which, as we shall see, it ultimately emerged triumphant. Thenceforward ' Experiment ', not ' Submit ' was the guiding principle of seekers after knowledge. The Moslem Orient, on the other hand, succumbed to the ordeal and so the supremacy in matters of scholarship gradually passed to the Christian Occident.

[1] *Vide* T. J. de Boer : *op. cit.*

CHAPTER IV

THE RENAISSANCE OF GEOGRAPHICAL STUDIES IN CHRISTENDOM

IF the intellectual task of the Dark Ages was to save what it could from the wreckage of ancient learning, that of the first succeeding centuries was to master what had been recovered and adapt it to the requirements of the Christian faith. From the tenth century onward we may watch this process at work, and there the intellectual revival of the Middle Ages may be said to begin. This revival was so remarkable that it behoves us briefly to inquire into its origin. Until recently it was popularly supposed that the Crusades were the main single factor. It was argued that these expeditions, by bringing the men engaging in them into touch with new, and, in a sense, more ancient civilizations, possessing different outlooks and traditions, provided at once a challenge to the accepted canons of Western scholarship and a stimulus to the intellect. Nowadays, however, it is generally recognized that a distinction must be drawn between the Crusades and the Age of the Crusades. The Crusades were only one expression, and that not the earliest—for the Renaissance had begun before the first Crusade—of the life of a vigorous new epoch. They brought East and West into closer contact, stimulated trade, transportation, and the use of money, and helped to accelerate many tendencies already at work. But that was almost the full extent of their influence, for their specifically intellectual consequences were less apparent, if indeed they are not largely hypothetical. Nor should this greatly surprise us. The Crusaders were fighters, not scholars. They were in no mood to carry on intellectual traffic with the Saracens, their enemies and the enemies of all that stood, in their view, for culture. Nevertheless, cultural relations with the Arabs of Syria were not wholly lacking. Early in the twelfth century Adelard of Bath is known to have visited Antioch and Tarsus, though it is not clear to what extent his acquaintance with Arabic science was gained there, while, towards the close of the Crusading epoch, Frederick II of Sicily included the Levant in the distribution of his questionnaires, and, when in Syria, came into direct contact with Mohammedan philosophers and scientists.[1] While the Crusades, *per se*, may not have had very much direct influence on this renaissance, their indirect influence was very considerable. Through the Crusades Christendom came into contact with a civilization different from its own. New fields of study were suggested as, for example, the oriental languages, from which, of course, many words passed into the western tongues, just as new plants and fruits and handmade wares passed westward. Then again, the Crusading tales of valour and chivalry could not fail to inspire the literature of the time. The cycle of the Crusades enriched the store of narrative poetry, while the *Historia Transmarina* of William of Tyre is second to none among medieval histories. Nor must we forget the importance of the habit of foreign travel fostered by the Crusades.

It is true that foreign travel was nothing new in Christendom. Pilgrimages as well as diplomatic and religious missions had been made ever since the new era came in. None the less, with the steady improvement in the means of communication, in the safety of travel and in the government of Europe, it came to assume a much greater significance about this time.

Now all the great religions, Judaism, Buddhism and Christianity, as well as Islam, have always had this in common, that while they were universal in scope, some places on earth were particularly sacred to them. To visit these places was the ambition of every devout follower. The Christian,

[1] *Vide* C. H. Haskins: *Studies in Mediaeval Science*, p. 130.

unlike the Moslem, was not obliged to go on pilgrimage, but all the same he was given ample inducements to do so in the guise of plenary absolution and indulgences. Rome, Santiago de Compostella, Tours, Chartres, Canterbury, Assisi, as well as Jerusalem, for the few who could afford it, were all centres of attraction. The popularity of these pilgrimages is clearly shown by the creation of hospices along the regular routes at such places as Roncesvalles in the Pyrenees and at the top of the St. Bernard and Septimer passes in the Alps.

In some cases, no doubt, the immediate motive for travel was not religious, but a desire to gratify the spirit of *Wanderlust*, or simply to 'see the world'. Many a knight who won honour and renown in Palestine had more of the pirate than the saint in him. But we are concerned less with their motives than their movements. In following these we are unfortunately hampered by the fact that the pilgrims, always on the alert for the miraculous and the mysterious, did not often trouble to record systematically either what they saw or where they went. Consequently, their itineraries have little geographical value, and we shall therefore treat them with scant courtesy.

The most notable records of Christian pilgrimages to Palestine are those of John of Würzburg, Joannes Phocas, Richard Cœur de Lion, Sigurd of Norway—all in the twelfth century, and that of Burchard of Mount Sion in the thirteenth century. A comparison of the following extract from John of Würzburg's *Description of the Holy Land* (he did not describe his journey thither) with any of the contemporary Arabic documents already described will suffice to impress upon us the superiority of the latter. It commences thus:

> Now because our redemption was begun in the city of Nazareth through the incarnation of our Lord, whereof annunciation was made by an angel, we propose to begin our description with this same city which is about sixty miles distant from Jerusalem and to touch briefly and compendiously upon the places which lie between it and the Holy City . . . this same city, which is ten

miles distant from Tiberias, is the chief town of Galilee and is properly called 'The City of the Saviour' because He was conceived and brought up within it: wherefore He was called a Nazarene. Nazareth is, being interpreted, a 'flower' or a 'shrub' and is justly so-named, because therein grew the flower with whose fragrance the whole world is filled. . . . In Nazareth runs that little fountain (legendary ?) from which Jesus in childhood was wont to draw water and take it to his mother. A mile from Nazareth to the southward is the place which is called 'the precipice' down which those who found Jesus wished to cast him, and to this day it is commonly 'The Lord's Leap'. At the second milestone from Nazareth is the city of Sepphoris, on the road which leads to Accon. Anna, the mother of Mary, who was the mother of our Lord, came from Sepphoris . . .[1]

And so on, until the writer reaches Jerusalem. Here he describes 'the holy places in the Sacred City, starting from the Church of the Holy Sepulchre and going round about, through the Gate of David', till he returns to the same place. Throughout, the narrative is devoted to the description of biblical topics and topography, and is singularly devoid of eye-witness information. There is scarcely a sentence that could not have been borrowed from existing literature or based upon the interpretation of some road-map.

The contemporary pilgrim narrative of Joannes Phocas is similar, but on the whole, more secular in outlook—witness the following description of Lebanon:

Mount Libanus . . . [is] a very great mountain clad in a robe of snow, hanging from it even as ringlets, overgrown with pine, cedar and cypress trees and adorned with numerous fruit-bearing trees of various kinds. The side thereof next the sea is inhabited by Christians, while the Saracens dwell on the side that looks towards Damascus and Arabia. From its ravines and hollows many rivers gush forth into the sea, beauteous and excessively cold at the time when the snow is melting, and chills the streams which feed them. At the foot of this mountain is Tripolis, which was built by its founder upon a peninsula; for a small spur,

[1] *Description of the Holy Land* . . . Edited by A. Stewart (Palestine Pilgrims' Text Society, 1890), pp. 3–4.

branching out from Libanus, runs out into the sea in the shape of a tongue, rising high at its eastern end. Upon the summit of this rising ground the builder of the city laid its foundations . . .[1]

This at least makes more interesting reading than John of Würzburg's narrative and it has the added virtue of being refreshingly geographical in content.

More effective than the pilgrim literature, and perhaps even more effective than the pilgrimages themselves, in remoulding current thought were the travels of scholars in search of learning. One of the distinguishing features of Western Europe about this time was an increased mobility in all classes of society.

No kind of men were more quickly touched by the new mobility than the thousands of youthful learners who desired to extend their knowledge, or, in some definite field, perfect their education. In the eleventh century such would commonly have sought a monastery, near or far. In the twelfth and then in the thirteenth century they followed the human current to the cities, where knowledge flourished as well as trade, and tolerable accommodation might be had for teachers and students.[2]

Certain towns, some for more, some for less obvious reasons, became homes of study. Bologna, Paris and Oxford are the chief examples. From far and wide, those who wanted to study Roman law came to Bologna, those who sought to read philosophy and theology were drawn to Paris (e.g. Abelard, Hugo, Peter Lombard, John of Salisbury, Albertus, Thomas Aquinas, Roger Bacon). 'What first brought masters and students to Oxford . . . is not so clear. . . . One hypothesis is that it was a colony from Paris, shaken off by some academic or political disturbance'.[3]

Finally, in assessing the factors contributing to the energetic human advance of western civilization at this epoch, we must not forget the significance of the re-conquest of central

[1] *Brief Description of the Castles and Cities from Antioch to Jerusalem. . . .* Edited by A. Stewart (Palestine Pilgrims Text Society, 1889), p. 9.

[2] H. O. Taylor: *The Mediaeval Mind*, Vol. II, p. 410.

[3] *Ibid.*: Vol. II, p. 416.

Spain from the Moors in the eleventh and twelfth centuries, and the contemporary occupation of Sicily and Southern Italy by the Normans. More than any other happenings, these paved the way for the acquisition of Arabic and Greek learning by Christian scholars.

Now, as we have already remarked, Western European scholarship in general was crippled down to this time by its ignorance of the Greek language. Even the alphabet was lost, with the result that at the hands of the medieval scribe Greek characters are gibberish, the word *grecum* commonly being inserted in its place—it was 'all Greek' to him. As late as the twelfth century we find the leading humanist, John of Salisbury, never quoting a Greek author unless he was available in a Latin translation.[1] In Eastern Europe and in Sicily the situation was different. There, with Constantinople and Palermo as their intellectual foci, the Greek traditions lived on. The Greek language remained the medium of law and government, civil as well as ecclesiastical, and the language of literature. And as the Byzantine Church was, at that time, well established in Asia Minor and Syria (whither Constantinople was drawn by strong geographical affinities), Greek culture enjoyed a wide vogue in the Levant, long after the Greek language had ceased to be the *lingua franca*. Here many of the 'classics' were translated into Arabic and Aramaean, sometimes directly, but more commonly by way of Syriac and Hebrew. With the territorial expansion of Islam along the whole of the eastern, southern and south-western seaboards of the Mediterranean went the rudiments of Arabic culture, to such good effect, in the case of Arabic letters, that, by the eleventh century, the libraries of Palermo and Toledo were almost as Arabic in language as those of Alexandria and Antioch. In composition, they consisted very

[1] The study of Greek was fostered in the Irish monasteries for a while. John Scot Eriugena, whose acquirements included some knowledge of Greek, was educated in Ireland, and according to Matthew Paris, Grosseteste was aided by a Greek priest of St. Albans in translating the *Testament of the Twelve Patriarchs* into Latin.

largely of commentaries on the classics rather than translations of them.

Prior to the twelfth century the intellectual contacts between Christendom and Islam were few and fitful. Gerbert [1] in the tenth century, and Constantinus Africanus in the eleventh, are conspicuous as students of Arabic scholarship more for their solitariness than for their accomplishments.[2] With the steady recovery of the lost European territory, especially of Moorish Spain, the Latin world gradually began to appropriate this neglected legacy. Toledo had become part of the kingdom of Castile in 1085. When the town was taken by Alfonso VI, it was made the capital of his kingdom. The Moslem inhabitants of the city soon found themselves living side by side with 'infidel' Christians, and it was not long before their presence in the same city as the King, the royal court and the Primate made a deep impression on the new-comers, who began to take a lively interest in the intellectual life of Islam. Early in the twelfth century, Raymund, the reigning Archbishop, founded a college of translators there, and entrusted it with the task of preparing Latin translations of the most important Arabic works on philosophy and science. Once the fame of its libraries had spread north of the Pyrenees, it was not long before all roads led to Spain. Adelard of Bath, Plato of Tivoli, John of Spain, Robert of Chester, to name only a few, played leading parts in this migration. Tired of editing and re-editing the aphorisms of the Church Fathers, they set about acquiring a working knowledge of Arabic and then applied themselves to the task of translation. Some settled down in Barcelona, others in Taragona, Leon, Segovia, and Pamplona, but Toledo was the chief attraction.

The most important meeting place of ancient and medieval culture in the twelfth century was, however, the Norman kingdom of Sicily, for only there did Greek, Latin and

[1] Gerbert was Pope (Sylvester II) from 999 to 1002.

[2] Thus Gerbert does not seem to have drawn on Arabic sources at all for the semi-geographical portions of his work *Liber de Astrolabio*.

Arabic civilization live side by side in peace and mutual toleration. The three languages were in current use, both for tongue and pen, so that knowledge of more than one of them was a necessity for the officials of the Royal Court, where men of learning from every land were welcomed. The production of translations was almost inevitable in such a cosmopolitan atmosphere, especially as it was directly encouraged by the Sicilian kings, from Roger to Frederick II and Manfred, as part of their policy for the spread of culture.

Little is known of the great majority of these translators, domiciled in Spain and Sicily, beyond their names and the names of the books they translated. Fortunately, fairly full details are known of Gerard of Cremona who can be regarded, with some justice, as their doyen. From these details, furnished by his pupils, we learn that he went to Toledo in search of Ptolemy's *Almagest*, which he could not find among the Latins and knew of only from literary references. He not only found it, but discovered also a multitude of Arabic books in every field, and, pitying the poverty of the Latins, learned Arabic in order to translate them. His version of the *Almagest* bears the date 1175. Before his death, which occurred at Toledo in 1187 at the age of seventy-three, he had turned into Latin at least seventy-one books. This number of works is catalogued in his version of Galen's *Tegni*. Three of them are on logic, several are on mathematics, including Euclid's *Elements*, and many are devoted to astronomy and astrology, while his list of Aristotelian translations is only exceeded, for length, by his medical translations from the works of Galen and Hippocrates. It is no exaggeration to say that Gerard did more than any other single person to render Arabic lore available to the Latin-reading world.

In the following centuries his pioneer work was carried on by such men as Alfred the Englishman, Michael Scot and Herman the German.[1]

[1] Not all these translations were from Arabic. A few were from Greek into Latin, e.g. those of Robert Grosseteste and William of Moerbeke

So epoch-making was the introduction of Arabic studies into the West that Renan [1] maintained that it divides the history of science and philosophy in the Middle Ages into two distinct periods. In the first, the human mind had, in order to satisfy its curiosity, only the meagre fragments of the Roman schools heaped together in the compilations of Martianus Capella, Bede, Isidore and certain technical treatises whose wide circulation saved them from oblivion. In the second, ancient science came back once more to the west, but this time more fully in the Arabic commentaries or the original works of Greek science for which the Romans had substituted compends.[2]

Yet this recovery of ancient learning could not have operated as it did but for the antecedent intellectual development. Before the first half of the twelfth century had passed, the patristic material had been mastered, along with the current notions of ancient philosophy, for the most part contained in it. The assimilation of this classical lore supplemented by what the Arabs had gained both from the Orient and their own observation led, in the twelfth and thirteenth centuries, to a quickening of the scientific impulse without parallel in the earlier period. There are manifestations of it at every turn. Adelard of Bath, for instance, blessed, it is true, with an inquisitive turn of mind, early found himself becoming deeply interested in what he was translating and it was not long before he was challenging many of the long-accepted principles of knowledge. Then we find him favouring scientific investigation rather than unquestioning faith, and setting reason above authority. In his *Natural*

who translated a number of the works of Aristotle at the instance of Thomas Aquinas (*c.* 1263), in order to make good the inadequacy of the Arabic translations of Aristotle. *Vide* F. Ueberweg : *Grundriss der Gesch. der Philosophie*, 2 Teil (11th ed., 1928), pp. 348 *et seq.*

[1] *Averroès et l'Averroïsme :* p. 200.

[2] 'Throughout all the mediaeval centuries, the "commentary" tended to supplant the work commented on, whether that work was Holy Scripture or a treatise on Aristotle.'—H. O. Taylor : *op. cit.*, Vol. II, p. 420.

Questions, one of two important original works of his, he tells his hypothetical questioner

I learned from my Arabian masters under the leading of reason ; you, however, captivated by the appearance of authority, follow your halter. For, what else should authority be called than a halter ? Just as brute beasts are led where one wills by a halter, so the authority of past writers leads not a few of you into danger, held and bound as you are by bestial credulity.[1]

In this same work, we find Adelard discussing questions which, for their diversity and sanity, give the book an extraordinary air of modernity. The following selection illustrates something of their scope : Why is the earth suspended in mid-air, and how is it maintained ? Why is sea-water salt ? What is the cause of earthquakes ? thunder ? lightning ? Why do winds travel along the earth's surface and not in an upward direction ? If the earth were pierced, how far would a stone, thrown down the hole, fall ? How do rivers succeed in maintaining a constant flow ?

While few of the questions asked by him were original,[2] no small credit attaches to him for his cool, unprejudiced treatment of them. He has little of the vested interest of the earlier scholars in orthodox biblical scholarship. At the same time he is no novelty-monger, for in one passage of another work of his—*De Eodem et Diverso*—he speaks impatiently and indeed contemptuously of ' those who harass our ears with daily novelties ' and of the ' new Platos and Aristotles to whom each day gives birth, who with unblushing front proclaim alike, things which they know and of which they know nothing and whose supreme trust is in excessive verbosity '.[3] For us his own work is important mainly as exemplifying another phase of the intellectual awakening and as revealing a hitherto almost unsuspected awareness of the importance of investigation into the cause of natural phenomena.

[1] *Natural Questions :* cap. 6.

[2] Cf. Aristotle, *De Caelo* and *Meteorologica, passim.*

[3] *De Eodem et Diverso :* p. 7 (H. Willner's Edition).

William of Conches, a contemporary of Adelard's, is important for much the same reason. He is essentially an apostle of reason in matters of natural science. To him the argument of his obscurantist co-religionists that 'we do not know how such a thing can be but we know that God can do it' calls forth the contemptuous answer that 'God can make a cow out of a tree, but has He ever done so? Therefore show some reason why a thing is so, or cease to hold that it is so.'[1] It is not difficult to see why the attempt to put this precept into practice should have brought William into conflict with the ecclesiastical hierarchy. Under pressure he withdrew some of his more heterodox contentions, such as his rejection of a part of the Genesis account of the Creation, including the miraculous birth of the first woman, but he never seems to have given up his views concerning the superiority of reason over authority in matters of scientific speculation. In spite of the clerical strictures his work *Dragmaticon*[2] enjoyed a wide circulation, as is attested by the number of manuscripts of it at Vienna, Munich, Paris, Canterbury, Oxford and other European centres of learning. It was, moreover, frequently cited by writers in the next century.

The inspiration of this new intellectual movement did not immediately permeate the geographical literature of the period. We shall look in vain for radical changes either in the content or in the character of the twelfth-century *liber cosmographicus*. The *Liber Floridus* of Lambert of St. Omer and the anonymous *De Imagine Mundi* (*c.* 1100) are both in the orthodox tradition. The author of the latter treatise states that he has placed nothing in his work 'except that which is approved by the best authorities' who, needless to say, are men like Isidore and Orosius. Lambert acted upon the same principle.[3]

[1] *Philosophia*: p. 29 (edition of 1531.).

[2] The name given to the revised version of the *Philosophia*.

[3] The descriptive—topographical and historical—works of Giraldus Cambrensis are in a class apart (*vide infra*, chap. vii). They owe far more to eye-witness observation than any other work—English or Continental—of the twelfth century.

By the thirteenth century the new leaven of experimental method was visibly working upon the old lump of patristic learning, and, although it would be untrue to say that the latter was transformed by it, a perusal of the geographical writings of such men as Thomas of Cantimpré, Bartholomew of England (Anglicus),[1] Robert Grosseteste,[2] John of Holywood (Sacrobosco), Vincent of Beauvais, Albertus Magnus and Roger Bacon cannot fail to convince us that an intellectual renaissance is in progress.

The *Tractatus de Sphaera* of John of Holywood (Joannes Sacrobosco, as he styled himself) reflects the changing times as well as any work coming from earlier decades of the century. It is extremely slight, comprising four chapters, which treat respectively of the terrestrial globe, of circles great and small, of the rising and setting of the stars and of the orbits and movements of the planets. Moreover, it is an entirely unoriginal work, adding nothing to Ptolemy's *Almagest* and his Arabic commentators. But by reason of its very simplicity and respect for authority, new and old, it achieved a remarkable popularity during the later years of the Middle Ages, and was still studied as the classic textbook on the subject long after Barozzi in 1570 had pointed out its numerous errors. It ran through no fewer than twenty-five printed editions before 1500, and another forty editions appeared before it was finally voted to be out of date.

Equally popular, but very far removed from it in size and content is the work (*c.* 1250) of Vincent of Beauvais. Finding the multitude of books too great for the ordinary reader, this Dominican Friar decided to collect certain extracts from the existing body of literature and classify them under the three headings of *Speculum Naturale*, *Speculum Doctrinale* and *Speculum Historiale*, treating respectively of natural science, philosophy and history. The result is an encyclopaedic work of a size hitherto unparalleled. It fills four large folio volumes in the early printed editions as against one octavo volume of Isidore's *Etymologiae*. Almost half of the

[1] *Vide* chap. vii. [2] *Loc. cit., passim.*

work, amounting to some 3,000 folio pages, deals with scientific topics. It is essentially an encyclopaedia culled from existing encyclopaedias. In the words of his *generalis prologus* the work is a collection of

certain flowers gathered according to my modicum of faculty from every one I have been able to read, whether of our Catholic doctors or the Gentile philosophers and poets. Especially have I drawn from them what seemed to pertain either to the building up of our dogma, or to moral instruction, or to the incitement of charity, or to the mystic explanation of its truth.

At the same time, the *Speculum* draws on more sources than any of its predecessors. Besides the ever-present Isidore and Pliny, the newly recovered Aristotle, many Arabic texts, especially on medicine (e.g. Avicenna and Averroès), and, significantly, quite a number of more recent writers, such as Adelard of Bath and William of Conches, are cited. An even more significant sign of the times is the fact that his geographical data on Asia include the contemporary narrative of John of Plano Carpini and Simon of St. Quentin. The arrangement of the *Speculum* is, however, in the accepted tradition, not only in its minutely scholastic classification and subdivision—the *Speculum Naturale* alone is divided into 32 books of 3,718 chapters—but in the subordination of everything to the biblical sequence of events. Thus the *Speculum Naturale* treats of natural science in accordance with the order of the six days' Creation; the *Speculum Doctrinale* starts with the fall of man and discusses the various forms of philosophy by which he may rise again; the *Speculum Historiale* deals with the story of Adam and his descendants.[1]

The topographical chapters of the *Speculum Naturale*, with the exception of those on Tartary, are thoroughly ' historical ',

[1] After considering how to proceed and group his matter, Vincent came to the conclusion that there was no better method than the one he has chosen, ' to wit, that after the order of Holy Scripture, I should treat first of the Creator, next of the Creation, then of man's fall and reparation, and then of events chronologically '.

consisting of fragments from earlier medieval writers. Frequently Vincent does nothing more than excerpt passages from his authorities. Thus, in speaking of the Ocean,[1] he first of all cites Aristotle's opinion, then Isidore's, that of the *De Imagine Mundi* and finally Adelard's. Similarly his treatment of the River Nile is simply a compilation of the relevant passages from the works of Isidore, Solinus, Comestor, Plato, *Deuteronomy*, Avicenna and Seneca, in that order.

Here and there the *Speculum* savours rather strongly of plagiarism, whole sections being taken from the works of previous compilers like Bartholomew of England and Thomas of Cantimpré, while a large portion consists of verbal repetition of Pliny's *Natural History*. On the other hand, his work does not include everything that is in previous medieval writers on nature, and the assumption made by many specialists in the study of medieval culture, that the *Speculum Naturale* is an adequate reflection of medieval natural science and that Vincent is sure to mention any previous writer or treatise, is far from true. 'His *Mirror* is a glass through which we see darkly and not face to face.'[2]

In the almost contemporary work of that great Dominican Friar,[3] Albert the Great (Albertus Magnus),[4] there is much more originality. We are assured by one writer that he is original even when he seems to copy.[5] The spirit of inquiry

[1] *Spec. Nat.*: V, 13.

[2] Lynn Thorndike: *History of Magic and Experimental Science*, Vol. II, p. 476.

[3] The Dominican Order had only recently been founded (1215), that is at the time when the western world was acquiring the new Aristotelian knowledge. Although primarily a religious movement (the advancement of *sacra doctrina*, the saving knowledge of Scripture, was the ultimate end of all study with Dominicans and Franciscans), the spiritual force animating it stimulated also the mental energies of those working within it with the result that the greater part of the intellectual output of the thirteenth century originated with members of these two mendicant orders.

[4] b. *c.* A.D. 1193–d. 1280.

[5] C. Jessen: 'Alberti magni historia animalium', *Archiv f. Naturgeschichte*, Vol. XXXIII, 1867, p. 99. Jessen spoke truly; for in copying Orosius, Albert makes his authority's description of the Spanish peninsula

is also more generally active. Thus, as a commentator of Aristotle, Albert is quite ready to question the validity of his master's arguments. 'I think that Aristotle must have spoken from the opinions of his predecessors and not from the truth of demonstration or experiment.' 'He who believes that Aristotle was God, ought to believe that he never erred. If one regards him as man, then surely he may err as well as we.'[1] In sentences such as these incidentally, Albert adumbrates the principles which informed his mind. As he puts it in another place: 'It is not the business of natural science to accept implicitly whatever may have been said, but to search for the causes proper to each natural thing. . . . Experience is the best teacher of all such things.'[2] His very chapter headings frequently suggest an antithesis between ancient authority and modern experimental method, e.g., 'Concerning the Iris of the Moon and what the Ancients have said of it and what the Moderns have tested by experiment'.

Albertus lived in an eager age; it was importunate for the enlargement of its knowledge. At the same time, it badly needed an interpreter, one who would attempt to expound the newly-acquired knowledge and fit it into the scholastic Christian scheme. This was Albertus' great accomplishment. 'It is our intention', he wrote at the beginning of his exposition of Aristotle's *Physics*, 'to make all the said parts intelligible to the Latins'. And he did it, not only for this single work, but for many others as well. Herein lies the main value of his work. On the purely experimental side, he achieved very little; even his own personal observations of natural phenomena do not seem to have led to novel conclusions.[3] As for the cosmographical part of his work

apply to the whole of Europe. Cf. *De Natura Locorum*: III, 7 and *Historia*: I, 2.

[1] *Physics*: VIII, 1, xiv.

[2] 'oportet experimentum non in uno modo, sed secundum omnes circumstantias probare'.—*Ethics*: VI, 2, xxv.

[3] *Vide infra*, chap. vii.

(contained in *Tractatus* III of *De Natura Locorum*), it tells us little that is new, and its statistics, impressive in their verisimilitude, merely deceive.[1] It is when confronted with controversy that Albert is seen to best advantage. Thus, in discussing the question of the habitability of the Southern Hemisphere, a subject which never failed to appeal to the medieval scholar or to reveal his intellectual calibre, Albert shows an unusually clear perception of the problems involved. Having satisfied himself that there is ample evidence, such as the presence of cities in southern India and of Ethiopian tribes near the Equator, for believing that the region 'which seemed to the Ancients to be a Torrid Zone is habitable',[2] he proceeds to investigate the question of 'whether the quarter of the earth which stretches from the Equator to the South Pole is habitable'.[3] He considers, first of all, the arguments against the theory. His chief opponent amongst the Ancients is, of course, Ptolemy. The arguments which Albert attributes to him are threefold.

> One of these is the straightness of the course of the sun round the [southern] tropic in its approach to Sagittarius and in its progress in Capricorn. The second cause is the perpendicular fall of the sun's rays on those parts . . . and the third is the proximity of the mass of the sun. . . .

He disallows the validity of these arguments, countering them with the Aristotelian view that between the heat of the equatorial zone and the cold of the southern polar zone there must be a mean temperate zone which is habitable as in the northern hemisphere. Then, assuming that there must be inhabitants in a habitable region—'for if it can be inhabited, then it would be vain unless it were inhabited by plants, animals and men'[4]—he attempts to anticipate the oft-raised

[1] E.g., in chapter iv we are told that the southern quarter of the habitable earth comprises 'two seas, eighteen islands, six mountain ranges, twelve provinces, forty-four cities, two rivers, besides several tribes of people'.

[2] *De Natura Locorum*: I, 6.

[3] *Ibid.*: *cap.* 7.

[4] Cf. Aristotle's principle that 'Nature does nothing in vain'.

objection that there is no first-hand evidence of such men ever communicating with those of the northern hemisphere, by suggesting two possibilities. In the southern part of the Torrid zone (according to 'a certain philosopher' Albert had read) there are mountains of rock which is said to be magnetic. The quality of the magnetism is such that it holds back flesh and blood (*carnes humanas*) just as the ordinary magnet holds iron. It is on this account that no one can cross them.

But [Albert continues] I do not think that this hindrance is operative everywhere, but only in certain places. Now in the survey of the world carried out under Caesar Augustus, we read that Augustus sent messengers to the kings of Egypt and Ethiopia who were to secure ships and the expenses necessary for those whom he had sent to make the expedition. When they came near the Equator, they found marshy places into which the Nile flowed and stony places which they could cross neither by ship nor on foot. They therefore returned without accomplishing their errand. I myself have seen some men who give as a reason why the mountains are not crossed the barren and precipitous nature of them, but it seems incredible to me that they should be impassable everywhere. I believe it is truer to say that the crossing is difficult but not impossible, and this, on account of vast deserts of sand rendered waste by the fierceness of the sun ; this makes the undertaking impossible unless provision is made for a long journey. It is for this reason, I think, that there is little or no intercourse between the men who live beyond the Equator in southern climes and those who live with us in the northern hemisphere.[1]

In his critical attitude towards authority and in his wide and discriminating use of sources of information, the Dominican Albert the Great is equalled only by the Franciscan Roger Bacon. Although the two men do not seem ever to have had any direct contact, they had much in common intellectually, and this notwithstanding Bacon's contempt of the rival order. In his *Opus Tertium* he speaks of a great man in Paris (Albert ? Thomas Aquinas ?) who is set up as an authority in the schools,

[1] *De Natura Locorum :* I, 7.

like Averroes or Aristotle, but whose works merely display 'infinite puerile vanity . . . superfluous verbiage and the omission of the most needful parts of philosophy'.[1] Many of Bacon's judgements on the fallibility of the ancients might well have been delivered by Albert. Consider, in this connexion, the following appraisal of Aristotle's work found in the *Opus Majus*.

> Of all famous philosophers Aristotle alone, with his followers, is stamped with the approval of all wise men, since he organized the branches of philosophy, as far as was possible in his age; but nevertheless he did not reach the limit of wisdom. . . .[2]

Or again:

> Not the philosophers only, but even the sacred authors have been subject to some human infirmity in this respect, for they have retracted very many of their own statements. . . . Even Paul withstood Peter, as he himself confesses. Augustine finds fault with the views of Jerome: and Jerome contradicts Augustine in many places. . . . Since, then, the facts are as stated, we should not give adherence to all that we have heard and read, but we should examine most carefully, the philosophical statements of older writers, that we may make up their deficiencies and correct their errors.[3]

In view of this we are not surprised to find that Bacon shares Albert's willingness to depart from the narrow path of patristic orthodoxy when the occasion seems to demand it. Here, for example, is what Bacon says of the extent and geographical position of the habitable earth.

> Ptolemy, in his book on the *Arrangement of the Sphere*, maintains that about a sixth part of the earth is habitable, due to the water, and that all the rest is covered by water. Therefore in the *Almagest* in the second book, he states that habitation is not known except in a quarter of the earth, namely, in that in which we dwell; whose length is from east to west and is one half of the

[1] Brewer's Edition, p. 30.
[2] *Opus Majus* (R. B. Burke's translation): Vol. I, p. 10.
[3] *Ibid.*, Vol. I, pp. 16–17.

equinoctial circle (i.e. equator) and its width is from the equinoctial circle to the pole. . . . But Aristotle maintains at the end of the second book of the *Heavens and the World* (i.e. *De Caelo et Mundo*) that more than a fourth is inhabited. And Averroes confirms this. Aristotle says that the sea is small between the end of Spain on the west and the beginning of India on the east. Seneca in the fifth book on *Natural History*[1] says that this sea is navigable in a very few days if the wind is favourable. And Pliny teaches in his *Natural History* that it was navigated from the Arabic Gulf to Cadiz . . . therefore the width of the earth through which the Red Sea[2] extends is very great; from which fact it is clear that the beginning of India in the east is far distant from us and from Spain, since the distance is so great from the beginning of Arabia toward India. From the end of Spain beneath the earth the sea is so small that it cannot cover three-quarters of the earth. This fact is proved by the weight of another consideration. For Esdras states in the fourth book that six parts of the earth are habitable and the seventh is covered by water . . . therefore, according to these facts the extent of the habitable portion is great and what is covered by water must be small.[3]

In other words, Bacon accepts the specious 'continental', rather than the orthodox 'oceanic' theory of the distribution of land and water.[4]

Continuing, he argues that

toward the poles of the world the water must abound because those parts are cold, owing to their distance, but cold multiplies moisture, and therefore from pole to pole the water runs down into the body of the sea and extends for no great width between the end of Spain and the beginning of India, and is called the ocean.[5]

Now, although the acceptance of this unorthodox theory implied belief in the existence of land—as Bacon himself readily admits—south of the Equator, he is careful not to offend the susceptibilities of the Church by postulating an antipodal landmass. Indeed, he appears to be aware of the

[1] i.e. *Quaestiones Naturales*.
[2] i.e. Indian Ocean.
[3] *Opus Majus*, *op. cit.*, Vol. I, pp. 16–17.
[4] *Vide infra*, chap. vii (*Oceanography*).
[5] *op. cit.*: Vol. I, p. 312.

danger of departing too far from the path of orthodoxy, for he proceeds, painstakingly, to convince his readers that his view was not only essentially reasonable (obtaining support from several writers of Antiquity), but that it was even compatible with certain statements of the Fathers themselves. In speaking of the land in the southern hemisphere, he confesses that it

> has not been measured in our times, nor have we found it attested, as it should be, in the books of the ancients ; nor is this strange, since more than half of the quarter [of the earth] in which we are, is unknown to us. . . . Similarly, if we speak about the two other quarters, and consider natural processes in accordance with natural philosophy, we find that those quarters are not covered by water as the majority of mathematicians estimate. For since the Poles and regions near them are of the same remoteness from the sun and the planets in accordance with the relative position of the poles to the paths of the planets at the middle of the world, between the two tropics, there must, in accordance with these facts be equal dispositions of land and water in our quarter and in the quarter beyond the equinoctial circle toward the other pole . . . therefore in the other quarter beyond the equinoctial circle there must accordingly be much [land] that is habitable ; at least as far as the regions whose latitudes reach 66 degrees, just as is the case here.[1]

But how did Bacon know that these lands were habitable ? Pliny and Ptolemy among the ancients and Ambrose and Basil among the Fathers had said so.

> We discover [from Pliny] that there is habitation on the tropic of Capricorn or beyond it. For a region in India is called Pathalis with a very noted port, as he says, where the shadows fall only in the south ; therefore dwellers here have the sun always to the north. He states the same thing in the Sixth book [of the *Natural History*] respecting the island of Taprobane in India. . . . But not only the philosophers, but also the sacred writers like Ambrose, in the *Hexaemeron*, and Basil, agree in the matter of this diversity

[1] Bacon places 'the end of the islands of Scotland and the Kingdom of Norway' in 66° N.

of shadows. For in the fourth book Ambrose says 'There are people located to the south in this world we inhabit who are seen to cast their shadow to the south'. Because he speaks expressly of shadows to the south, he can be understood to have reference to those who have shadows falling to the south only, namely, dwellers on the tropic of Capricorn and beyond it; because the sun is always north of them except at one time when it is over their heads on the tropic of Capricorn.[1]

We see, then, that although Bacon occasionally parts company with the Fathers, he is far from being a rebel against ecclesiastical authority. On the contrary, he asserts categorically, at the beginning of the *Opus Majus*, that there is no fundamental opposition between pagan and Christian wisdom, for 'there is only one perfect wisdom . . . [and] from the roots of this wisdom all truth has sprung'.[2]

If Bacon shares the intellectual outlook of Albert, he is equally heir to his limitations. With the exception of a reference to the Rubruck expedition, his regional description of the world is essentially the same as Albert's, and, consequently, not very different from those of his classical forbears—and this notwithstanding his assurance that he went 'over sea and to various lands and to annual fairs',[3] in order that he might see the things of Nature with his own eyes. Here and there Bacon appears to have known less than some of the Ancients. Concerning Ethiopia, for instance, he knew considerably less

[1] *op. cit.*: Vol. I, pp. 325–7.

[2] *op. cit.*: Vol. I, p. 36. The fact that the *Opus Majus* was dedicated to Pope Clement IV, whose patronage Bacon hoped to secure, no doubt helps to account for his compromising attitude toward science and religion. 'He was obsessed with the [scholastic] idea that philosophy, including every branch of knowledge, must serve theology, and even in that service find its justification. But what has chemistry to do with theology? What has mathematics? And what has the physical experimental method? By maintaining the utility of these for theology, Bacon saved his medieval orthodoxy, and it may be, his skin, from the fire. But it wrecked the working of his genius.'—H. O. Taylor: *op. cit.*, Vol. II, p. 516.

[3] *Vide* F. A. Gasquet: 'An Unpublished Fragment of a work by Roger Bacon', *English Historical Review*, Vol. XII, p. 502.

than Ptolemy. We quote his account as showing the 'historical' nature of his geography.

Between the straits of the Red Sea and the Sea of Ethiopia,[1] Ethiopia begins. Where the latitude of the region is about sixteen degrees and the length of the longest day about thirteen hours according to Ptolemy in the *Almagest*—Pliny also is in practical agreement in his Sixth Book[2] and his second chapter—is Saba, an island formed by the Nile, the royal city of Ethiopia, concerning which we read in Isaiah 'the labour of Egypt, and merchandise of Ethiopia and of the Sabeans'.[3] This is Meröe which is at the remotest point in Ethiopia and at the end of the known habitable portion of the earth . . .[4] and it is mentioned in *Ezekiel* xxvii. . . . This city is distant from the sea of Ethiopia about 700 miles according to Pliny in the sixth book. It is situated upon the first climate from which the climate receives the name of Diameröes.[5] Candace, a woman, was ruler there.[6] . . . Where the latitude of the region is about the same on the shore of the Red Sea is the city of Ptolemais, founded by Ptolemaeus Philadelphus for the earliest hunting of elephants. . . . Then beyond, in the same latitude but towards the west, between Meröe and Ptolemais, 4820 stadia distant from Ptolemais, as Pliny states and Bede confirms . . . is Berenice, the city of the Ethiopian Troglodytes. Scripture mentions these Troglodytes in the second book of the Apocrypha. . . .[7]

According to Isidore, in the ninth book, there are three principal tribes of Ethiopians, the Hesperi, Garamantes and Indi. The Hesperi are in the west, the Garamantes in the middle, the Indi in the east. . . . There are, however, many other Ethiopians joined to these tribes in different places, much degraded also from that which human nature should be, whose names, localities and manners it is not within the scope of the present plea to give. . . .

Ethiopia terminates below at the Red Sea eastward, with Africa to the west and Egypt in the middle between these extremes. In this middle division is the city of Syene, of which Ezekiel speaks

[1] i.e. Indian Ocean. [2] i.e. of his *Natural History*. [3] chap. xlv, v. 14.

[4] cf. Ptolemy's 'Montes Barditi qui meridionalem limitem nostra habitabilis terminant' to which he gives the latitude 16° S.

[5] i.e. 'through Meröe'. [6] *Vide Acts* viii. 27.

[7] *Opus Majus* (R. B. Burke's translation): Vol. I, p. 331.

expressly in chapters xxix and xxx. . . . Syene is the lower limit of Ethiopia and the remotest part of Egypt . . . but Meröe is the upper[1] limit of the known inhabited portion of the earth, according to Pliny. . . . From Syene to Meröe according to Pliny . . . are 5,000 stadia. . . . Syene is located on the Tropic of Cancer.[2]

Fortunately, Bacon's fame as a geographer does not rest solely upon descriptions such as these. As with Albert it is his methods rather than his materials that are valuable. If we are disposed to complain that these methods are more critical than constructive, we must recollect that in medieval times there were plenty of wrong ideas to combat. It was not easy to find a firm foundation on which to build or materials without some hidden defect. So Roger Bacon and Albert the Great, by devoting much of their space to the collation and criticism of earlier works, avoided novel theories —a wise precaution for their times. While both men urged the value of personal experiment in scientific pursuits, neither of them appears to have been a great experimenter. One of Albertus' biographers declares : 'I do not know of his undertaking a single experiment in order to solve a physiological or physical problem.'[3] Of laboratory experiment, of scientific instruments, of exact measurements, they had no more notion apparently than their contemporaries. Consequently, to regard either of them as a pioneer of unfettered scientific research is to give credit where none is due. They were essentially exponents of scholasticism, standing firm for the truth of the Christian revelation, and seeking to reconcile all knowledge logically with the One Truth, as it was understood by the Church. They never attempted to exercise, nor did they permit others to exercise, that wider rationality which seeks to embrace the whole of human experience without such arbitrary boundaries as dogmas prescribed by authority. They wrote as clerics for clerics, to promote the welfare of the Church and Christianity, and their sole motive in writing

[1] i.e. southern. [2] *op. cit.:* Vol. I, pp. 331–2.

[3] *Vide* L. Thorndike : *History of Magic and Experimental Science*, Vol. II, pp. 536–7.

was to promote a sound and saving knowledge of God. For this reason their discussions do little more than confirm the evidences of a changing outlook on life and learning which we have obtained from other and earlier sources.

Neither man had many immediate followers. No doubt 'the paralysing mists of Scotian dialectic', which enveloped the study of Greek science in the following century, had something to do with it, as Bridges surmises,[1] while it may well be that the relatively belated publication of their works,[2] after the invention of the printing press, provides a contributory reason of some relevance. At the same time it would be an error to suppose that the life-work of Albert and Bacon was lost to posterity. Here and there throughout Europe the memory of it survived as a stimulating force, and kept the spirit of scientific inquiry alive until the Renaissance of the fifteenth century. We must never forget, of course, the 'link' between the *Opus Majus* and the discovery of America.[3] The arguments used by Bacon to establish the existence of habitable land at no great distance westward from the coast of Spain were repeated almost *verbatim* by Pierre d'Ailly, and, however uncertain the premises and unsound the conclusions may appear to a modern reader, they were seized on by Columbus and his contemporaries and stimulated exploration to the west and south. As Bacon's arguments reveal, King Solomon's Mines were located in the far distant S.E. extremity of India, and Columbus' belief that he had come within reach of them, first in Española and later in Darien, was not an absurd one in the light of current theory.[4]

English Renaissance geography also owed a great debt to Roger Bacon, not merely for his *Opus Majus*. His general

[1] *The 'Opus Majus' of Roger Bacon:* Vol. I, p. xxxiii.

[2] There are no *incunabula* of Bacon's work and comparatively few of Albert's. Even down to the end of the sixteenth century, the *Opus Majus* was still very scarce. Ortelius, writing to Camden in 1588, speaks of the latter having promised to let a certain man, J. Cole by name, make a copy of his specimen of Bacon's Geography for him.

[3] *Vide infra*, chap. ix.

[4] E. G. R. Taylor: *Scottish Geographical Mag.*, Vol. xlvii, pp. 218–19.

teaching as to the importance of applied mathematics and of the experimental approach to science profoundly influenced such men as Richard Eden and John Dee, that is, the pioneers of the English geographical Renaissance.[1]

Once the Greek and Arabic 'classics' had been translated into Latin, their further translation into the vernacular was only a matter of time. Latin, then as now, was an unknown tongue as far as the masses were concerned, and so the spread and popularization of knowledge was dependent upon oral communication and translation into the spoken languages of the day. In the ninth century King Alfred was lamenting that, although the

> Churches throughout the whole of England stood filled with treasures and books, and there was also a great multitude of God's servants, [yet] they had very little knowledge of books, because they could not understand anything of them because they were not written in their own language (i.e. Anglo-Saxon).[2]

It was because of this that he set about translating 'some books which are most needful for all men to know, into the language which we can all understand . . .'[3] In the same century Raban Maur, for all his Latin learning, kept his interest in his mother tongue and encouraged preaching in German. With the aid of his favourite pupil, Walafrid, he prepared German glosses and Latin–German glossaries for Scripture. Before his death popular translations of the Gospels had appeared. Then, in the last part of the tenth century and the first part of the eleventh, Notker the German,[4] a monk of St. Gall, made German translations of Boethius, Virgil, Terence, and Capella (*De Nuptiis*). And before the tenth century was out, Bede's *De Natura Rerum* was translated into Anglo-Saxon.[5]

[1] *Vide* E. G. R. Taylor: *Tudor Geography, passim.*

[2] Preface to Gregory's *Pastoral Care.*

[3] *Vide Essay on the State of Literature and Learning under the Anglo-Saxons,* T. Wright, p. 52 *et seq.*

[4] d. 1022.

[5] *Vide* Cotton MS. Tiberius, B.v. (British Museum.)

By the end of the twelfth century these languages were being increasingly employed, not only for speech, but for professional and diplomatic uses. But scholarship was still largely the monopoly of the clergy. As they continued to write in the language of their liturgies, they did nothing to promote the laicization of learning. Consequently, few *classical* works made an early appearance in vernacular tongues. However, several geographical works of a semi-popular character found their way into circulation before the end of the thirteenth century. Notable among these are the *Image du Monde*, commonly attributed to a certain Gossuin of Metz, and the Norwegian *Konungs Skuggsjá* (= The King's Mirror), of unknown authorship. The former, originally composed in metrical form, but almost immediately turned into prose, is the work of a well-read man. Many authors, both classical and medieval, are quoted and sometimes copied word for word. Notwithstanding, it is more than a compilation and contains sufficient of the grotesque [1] and unexpected to assure for itself a long-lived success. So popular did it prove, in fact, that it was soon re-written in prose form and translated into a number of other spoken tongues, including German and English.[2] The latter work of approximately the same date, viz. *c.* 1250–60, is geographically of much greater value. In spite of his remoteness the author was better informed than his French contemporary and he was obviously far more intelligent.

Although well acquainted with much of the earlier mediaeval literature, he has liberated himself to a surprising extent from its fabulous conceptions. We hear nothing of the many fabulous people who were still common amongst much later authors, nor about whirlpools, nor the curdled and dark sea, but instead we have fresh and copious information about the Northern regions . . .[3]

On subjects of special interest to northern dwellers, such as

[1] e.g. II, 5 *et seq.*, on the marvels of India.
[2] *Vide* W. Caxton's translation entitled *Mirrour of the World.*
[3] F. Nansen : *In Northern Mists*, Vol. ii, p. 243.

glaciers, icebergs and geysers (the author had lived in Iceland) the *King's Mirror* is without rival. He writes of Greenland in the following authentic terms:

> There are small portions of the land which are thawed, but all the rest is covered with ice, and the people do not know whether the country is large or small, since all the mountains and valleys are covered with ice, so that no one can find his way in. . . . Men have often tried to go up the country, upon the highest mountains in various places, to look around them, to see whether they could find any part that was thawed and habitable, but they have not found any such, except where people are now living, and that is but little along the shore itself. . . . The land is cold, and the glacier [1] has this nature, that he sends out cold gusts which drive away the showers from his face, and he usually keeps his head bare [of cloud?]. But his near neighbours have to suffer for it, in that all other lands which lie in the neighbourhood get much bad weather from him and all the cold blasts that he throws off fall upon them.[2]

In view of such descriptions as these, and they abound on the pages of the *Mirror*, we are hardly surprised to find Dr. Sarton hailing it as 'the most important geographical work of Christendom in the first half of the thirteenth century'.[3]

Remarkable as it is geographically, the *Konungs Skuggsjá* can hardly claim to be the most famous medieval vernacular writing, for none is so well known as that curious English document entitled *The Travels of Sir John Mandeville*. While it is probably true that all but a few pages of it were plagiarized from existing literary sources by a certain doctor, Jean d'Outremeux by name, and while the probability of the author's having seen with his own eyes all the countries he describes is very remote, the work has unquestionable value as illustrating the fourteenth-century layman's idea of the world.

> The earth [he tells us] is full large and full great, and holds in

[1] i.e. the ice-cap. [2] Nansen's translation, *op. cit.*, Vol. ii, p. 246.
[3] *Introduction to the History of Science*: Vol. II, part 1, p. 43.

roundness and about environ, by above and by beneath, 20,425 miles, after the opinion of old wise astronomers [1]; and their sayings I reprove nought. But, after my little wit, it seemeth me, saving their reverence, that it is more.

And for me to have better understanding I say thus. Be there imagined a figure that hath a great compass. And, about the point of the great compass that is clept the centre, be made another little compass. Then after, be the great compass devised by lines in many parts, and that all the lines meet at the centre. So, that in as many parts as the great compass shall be departed, in as many shall be departed the little, that is about the centre, albeit that the spaces be less. Now then, be the great compass represented for the firmament, and the little compass represented for the earth. Now then, the firmament is devised by astronomers in twelve signs, and every sign is devised in thirty degrees; that is, 360 degrees that the firmament hath above. Also, be the earth devised in as many parts as the firmament, and let every part answer to a degree of the firmament. And wit it well, after the authors of astronomy, 700 furlongs of earth answer to a degree of the firmament, and these be 87 miles and four furlongs. Now be that here multiplied by 360 sithes, and then they be 31,500 miles every of eight furlongs, after miles of our country.[2] So much hath the earth in roundness and of height environ, after mine opinion and mine understanding.[3]

[1] Actually, this figure does not agree with any of the reckonings of the 'old astronomers'. Ptolemy's is the nearest, 22,500 miles. It was only the 'new' (medieval) astronomers, such as Honorius of Autun and William of Conches, whose figures approximated to the one given, and even then the usual figure was 20,520. The discrepancy is, no doubt, due to the employment of a different value for the stadium, the basic unit of length in Antiquity, whose relation to the mile was variously interpreted. *Vide* E. G. R. Taylor: 'Some Notes on Early Ideas of the Form and Size of the Earth.' *Geographical Journal*, Vol. 85, pp. 85 *et seq.*

[2] This was Isidore's reckoning (*vide ante*, chap. ii, p. 24) and was based on Macrobius' (i.e. on Eratosthenes') assumption that the circumference of the earth was equal to 252,000 stadia. By equating stadia with furlongs, the value of 87½ miles, or 700 stadia, was accordingly obtained for the length of a degree.

[3] *The Travels of Sir John Mandeville: the version of the Cotton MS. in modern spelling* (Library of English Classics): pp. 123–4.

On the question of the Antipodes he affirms:

I have gone toward the parts meridional, that is, toward the South, and I have found that in Lybia men see first the star Antarctic. And so far I have gone more further in those countries that I have found that star more high; so that toward the High Lybia it is eighteen degrees of height and certain minutes (of the which sixty minutes make a degree). After going by sea and by land toward this country of that I have spoken, and to other isles and lands beyond that country, I have found the Star Antarctic of thirty-three degrees of height and more minutes. And if I had had company and shipping for to go more beyond, I trow well, in certain, that we should have seen all the roundness of the firmament all about. . . . I say you certainly that men may environ all the earth of the world, as well under as above, and turn again to his country, that had company and shipping and conduct. And always he should find men, lands and isles, as well as in this country. For ye wit well, that they that be toward the Antarctic, they be straight, feet against feet, of them that dwell under the Transmontane[1]; also well as we and they that dwell under us be feet against feet. For all the parts of sea and of land have their opposites, habitable or trepassable, and they of this half and beyond half . . .[2]

Strangely similar in its implications is the contemporary Latin work entitled *Directorium ad Passagium Transmarinum.*[3] Without attempting to substantiate or refute the claim it makes to be an authentic travel narrative, we may use it as revealing some of the more popular geographical notions of the time. Like the *Travels*, the *Directory* clearly postulates a habitable equatorial zone and a southern landmass of considerable proportions. Thus the writer speaks of having passed the equatorial regions into a Southern latitude where he no longer saw the arctic Pole Star, but in its place the 'Antarctic'

[1] i.e. the Pole Star.

[2] *The Travels of Sir John Mandeville: op. cit.*, pp. 121–2.

[3] Edited for the *American Historical Review*, 1907, by C. R. Beazley, who attributes the work to W. Adam, Dominican Bishop in Persia, 1323–9.

at an elevation of 24°, while he avers that certain merchants penetrated to a point where the elevation became 54°.[1]

As Beazley demonstrates in his edition of this work, it is not beyond the realm of possibility that the writer had been south of the Equator. Madagascar was already known through the writings of Marco Polo,[2] so that the author may conceivably have penetrated to a point in the South of that island where the Antarctic star would have an elevation of 24°.[3] Whether he did or not must ever be a matter of conjecture, but it does at least point to a belief in a traversable and habitable torrid zone. That the writer was himself out to prove this hypothesis is plain from another passage of the document, wherein he concludes that there must be more habitable land to the South and East than is commonly supposed, and that the concept of the Antipodes is neither frivolous nor false.

Exactly where fact ends and fiction begins in these narratives we have no means of telling; nor does it matter materially, seeing we are more concerned with popular beliefs than with the precise determination of world horizons. What is important is that stories of this kind exerted a profound influence upon the laity of the late Middle Ages. 'Travellers' tales' have ever made a wide appeal, for the very good reason, if for no other, that men universally have a natural tendency to assert, and a liking to hear, things that smack of the sensational, the extravagant and the incredible, and to fly in the face both of reason and experience. 'Lord, I believe, help Thou my unbelief,' expresses a common attitude of mind, in things material as well as spiritual, and in ancient and modern times alike. It requires little imagination therefore, to conjure up the impact of such tales as *The Travels of Sir John Mandeville* upon the minds of men living in an age that was supremely the age of faith. The reading of them, as also of the *Mirabilia* (or wonder books), the *Bestiaries*[4] (or animal books) and the

[1] *American Historical Review:* Vol. XII, p. 821.

[2] *Vide infra*, chap. vi. [3] *Amer. Hist. Rev.:* Vol. XIII, p. 67.

[4] *Vide The Bestiary* of Philippe de Thaun in T. Wright's *Popular Treatises on Science written during the Middle Ages.*

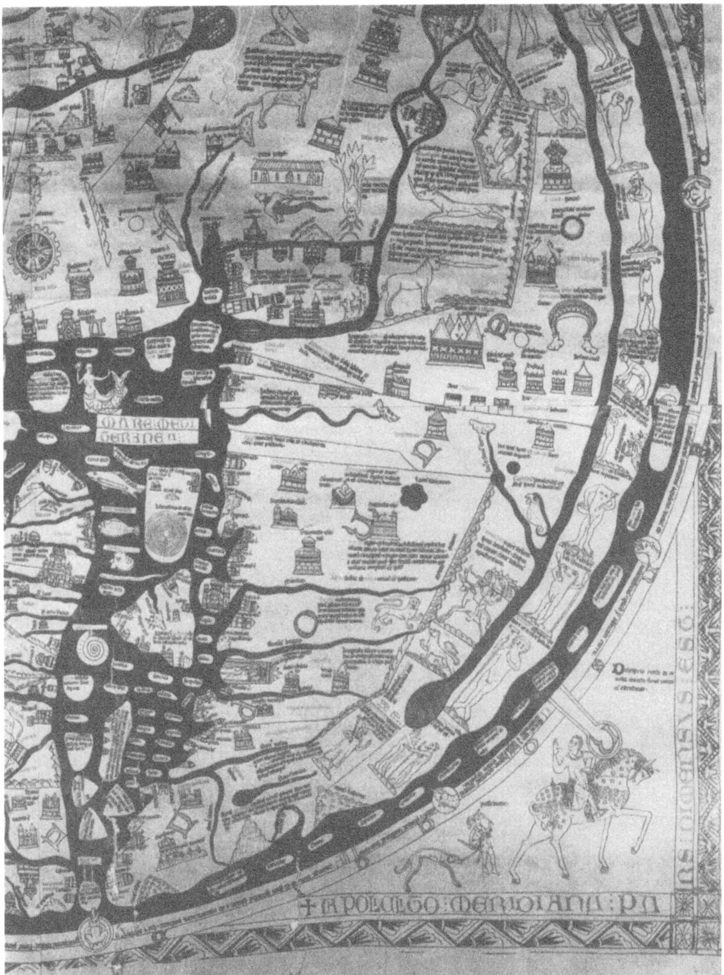

VI

AFRICA—ACCORDING TO THE HEREFORD WORLD MAP (c. 1280)

(*Hereford Cathedral*)

The reader will observe that, owing to an oversight on the part of the map-maker, Africa is called Europe.

Lapidaries (stories of the properties, mostly magical, of gems) thrilled not only the common people but men of education as well, and inspired the pictorial illustrations of many famous *mappaemundi.*[1] Indeed it would be difficult to over-emphasize the fascination of this kind of literature for the leaders of medieval and renaissance thought, whether in geography, natural history or zoology. As late as 1505, we find Duarte Pacheco, the great Portuguese navigator and scientist, speaking one moment with the exactitude of a modern ' pilot ' book and the next about the existence of mile-long serpents ! [2] When we recollect that Pliny's *Natural History*, Solinus' *Collectanea* . . . and the anonymous *Physiologos* (a collection of stories based upon the marvellous peculiarities of animals, composed c. A.D. 200, to which Christian allegories were subsequently added), were three of the chief source-books of the Middle Ages, the domination exercised by these teratological traditions is readily inderstood.[3]

[1] Notably the Hereford and Ebstorf maps, *vide* Plate No. VI.

[2] *Esmeraldo de Situ Orbis* : I, 27, (Hakluyt Society) edited by G. H. T. Kimble.

[3] Neither the *Physiologos* nor the ' wonder-books ' are entirely typical of the attitude of educated people in the late Middle Ages towards Nature. Thus, concerning the story of the pelican which restored its young with its own blood, and other similar fables found in the *Physiologos*, Albert writes that they belong to the realm of story books rather than to the realm of things ' proved philosophically by experience '. At the same time, he is far from rejecting wholesale all of Pliny's equally incredible assertions.

CHAPTER V

EXPANDING HORIZONS—AFRICA

THE 'travellers' tales' and wonder-books of the fiction monger paved the way for the even stranger stories of the explorer. The apathy of the scholar notwithstanding, these stories had a large currency and gradually succeeded in moulding geographical opinion, learned and popular alike.

It was from the South that the first reports of exploration came. Soon after the time of Al-Masʿūdī, Arabs in search of converts and commerce succeeded in establishing contact with the lands south of the Sahara. By 1076 the Sudanese kingdom of Ghana (now one of the lost cities of Africa) had come to acknowledge the suzerainty of the Moslem Empire. This rendered free movement across the Sahara possible and opened up the way for the commercial exploitation of the Negro lands. Many years elapsed, however, before the peoples of Christian Europe figured directly in this trade, partly from their anti-Moslem prejudice, and partly because any organized participation in that trade would have evoked the bitterest opposition of the Moor.

It was not until friendly intercourse with Moslems in Spain and Sicily during the twelfth century had revealed to Christian peoples the existence of a civilization in the interior of the continent, that they began to show any lively interest in the Sudan. Then arose the desire, above all, to penetrate into the country of Wangara, where, it was said, there was much gold.

The first indisputable indication of this interest dates from the year 1223 when the Genoese erected a factory at Tunis. Here they established commercial relations with traders from

Timbuktu. We may take it that first-hand European contact with the Sudan was established not long afterwards. The earliest authentic eye-witness account of the region, by a Christian, was given in the year 1283. It was then that Ramon Lull wrote in his *Blanquerna* about the messenger of a certain cardinal having undertaken a journey southwards across the Sahara. He

> found a caravan of 6,000 camels laden with salt leaving a town by name Tibalbert and going to that country wherein is the source of the River Damiata.[1] So many people did this messenger find there that in fifteen days all the salt was sold: and these men were black and adored idols. In that land is an island in the middle of a great lake and on that island there lives a dragon to whom all people of that island do sacrifice . . . and the people marvelled greatly at the messenger, because he was a white man and a Christian: for never before had they heard of any Christian coming to that land.[2]

Elsewhere in the same work we have a further indication that knowledge of the Sudan was then beginning to filter through to southern Europe. 'There entered [the Papal court] a Gentile[3] who came from a southern land which lies within the regions of the desert and from a city which is called Ghana. In that land there were many princes who adored idols and likewise adored the sun and stars and the birds and beasts.' The stranger goes on to relate how that in his quest for the true religion he journeyed through divers lands 'to inquire what that thing might be which was worthy of adoration above all things else', and how that eventually he had arrived at Rome. As a result of his mission, the Pope sent 'friars that had learnt Arabic, the which friars went to the Saracen king . . . and, by the grace of God, they converted the Saracen king and a great number of people'.[4]

It is highly tantalizing that the narrative should stop short

[1] i.e. River Nile.

[2] p. 373 (A. Peers' Edition).

[3] i.e. a heathen.

[4] *Blanquerna*: p. 356 *et seq.*

with such a brief notice of these lands, but it must be confessed that Lull's interests were avowedly not geographical. For the first exhaustive travel book on the Sudan we have to look to the next century and to an Arab, not to a European.

He was Ibn Abdallah Mohammed, surnamed Ibn Batūta—'the traveller not of an age, but of Islam'.[1] Born at Tangier at the beginning of the fourteenth century, he manifested his zeal for travel at an early age, for when only twenty-two he undertook the pilgrimage to Mecca, visiting at the same time the Arab colonies on the East African coast. In this, of course, he was merely following the trail of Masʿūdī.

After eighteen years of almost continuous journeying in China, Malay, Sumatra, Central Asia and, perhaps, in the Arctic, he returned in 1352 to his native land, only to set out shortly afterwards for the 'Nile of the Negroes'. His narrative, written presumably on his return from the Sudan, is rich in the lore of the times, and throws considerable light upon the question of the medieval knowledge of Africa. The following quotations are taken from a translation of an abridged Arabic manuscript preserved in Cambridge.

Having followed the regular caravan route across the Sahara from Sigilmesa to Tagaza, Batūta finds himself in a country where, owing to the exigencies of the climate, there is

> neither water, bird or tree, but only sand . . . blown about by the wind so that no vestige of a road remains. People can travel, therefore, only by guides from the merchants, of which there are many. . . . We passed it (i.e the desert) in ten days, thence to Abu Latin. . . . This is the first district of Sudan. . . . The native's clothing is all brought from Egypt. The greater part of the inhabitants are merchants. . . . I then proceeded . . . to Mali,[2] the distance of which is a journey of 24 days, made with effort. . . . After ten days from leaving Abu Latin we came to the village Zaghari which is large and inhabited by black merchants. Among these live a number of white people. . . . We

[1] C. R. Beazley: *The Dawn of Modern Geography*, Vol. III, p. 535.

[2] Melli of the Catalan Atlas and Viladestes Portolan chart, *vide* Plate No. VII.

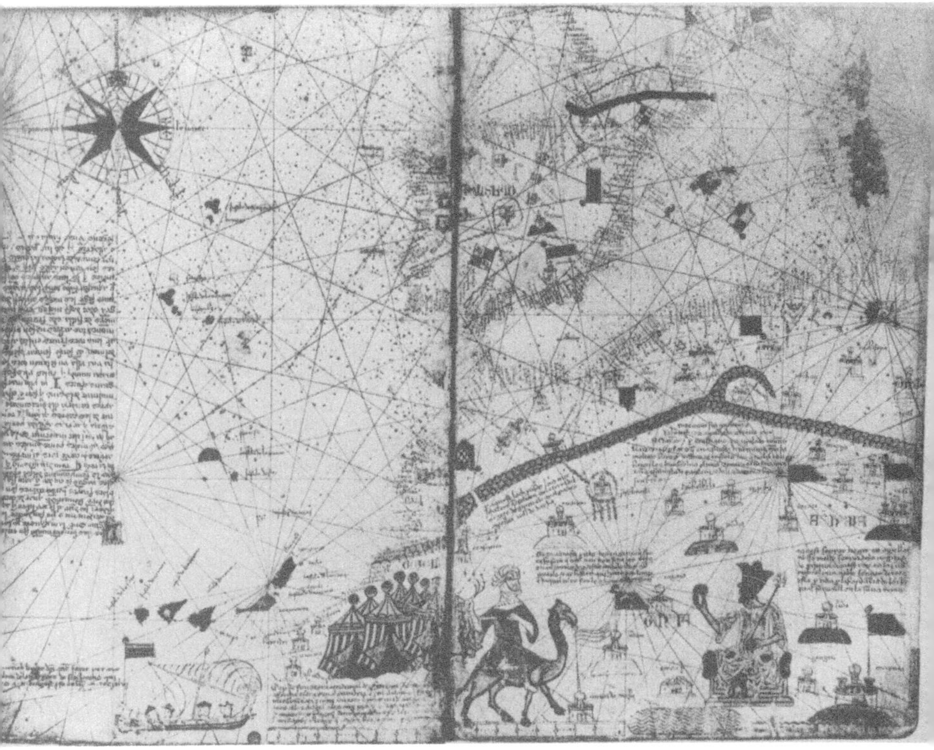

VII. AFRICA—ACCORDING TO THE CATALAN ATLAS OF CHARLES V (1375)
(Bibliothèque Nationale, Paris)

then left this place and came to the great river which is the Nile (i.e. Niger); upon it is the town of Karsanju from which the Nile descends to Kabara, then to Zaga, the inhabitants of which were the first to embrace Islam. . . . From this place the Nile descends to Tambucutu, then Kawkaw,[1] [and] proceeds to the town of Muli, which is the extreme district of Mali[2]; [it] goes on to Yuwi[3] the greatest district of Sudan . . . no white man can enter here. . . . The Nile then descends from this place to the countries of Nubia, the inhabitants of which are Christians, then to Dongola. . . . The Nile then descends to the Cataracts which terminate the regions of Sudan. . . .[4]

And so the narrative continues, carrying throughout unmistakable signs of its being the work of an eye-witness. The writer is content to describe what he has seen and seldom, if ever, allows his imagination to have free rein. Thus he steadily refuses to surmise what lies beyond the Niger, and so makes it impossible for us to envisage his idea of the continent as a whole. That his silence is to be construed as arising from ignorance is clear from a little passage dealing with the eastern coasts of Africa. In speaking of Kulwa (= Quiloa), Batūta says: 'A merchant told me that the town of Sofala is half a month's journey from Kulwa and one month's journey from Yuwi, in the country of Limiyin,[5] and that gold is brought from Yuwi to Sofala'.[6] While it is true that this statement is made in the name of a fellow-traveller, there is little reason for supposing that it was not Batūta's own view. At all events, he makes no criticism of it.

Three things, in the main, emerge from the reading of Batūta's African travels. In the first place it elicits the fact of extensive and far from inaccurate knowledge of the Sudan.

[1] Gaogao of the Catalan maps.

[2] The kingdom of Melli is believed to have extended south of the great northward bend of the Niger, from its source to its confluence with the Rima River near the modern town of Gomba.

[3] Probably Yeou, near Lake Tchad.

[4] *The Travels of Ibn Batūta*, translated by S. Lee: p. 233 *et seq.*

[5] i.e. Lemlem, a region of the upper Niger.

[6] Quoted by W. D. Cooley: *Negroland of the Arabs*, p. 93.

His theory that the Niger flows eastwards is a distinct advance upon that of Idrīsī and his school. Secondly, it demonstrates that there was commercial and religious traffic across the Sahara on a big scale at this time, and that north African and possibly European Moslems were actually in residence at the chief marts of the Sudan. Thirdly, that the negro converts to Islam were regularly undertaking the pilgrimage to Mecca.

At last, therefore, the agelong speculations concerning tropical Africa were being replaced by authentic accounts of civilization in a region which, by many, had been thought to be uninhabitable—a belief which was fondly held by the schoolmen long after this time. In view of the vitality of these Senegal-Niger peoples, it is a matter for some surprise that so little knowledge should be forthcoming about the regions beyond them. In none of the geographical writings of the Middle Ages is mention made of Arab or Christian contact overland with the Gulf of Guinea. This circumstance can hardly be fortuitous. The most obvious explanation is furnished by the climate and natural vegetation of the southern Sudan. Once in the savanna zone the camel ceased to be the 'ship' that suited so admirably the conditions of the desert, while the fly-infested swamps with their unwholesome miasmas effectively barred the progress of the unacclimatized Arab. Even so, it will be objected that these facts alone cannot explain the erroneous information that was in general circulation regarding these lands. So far were some travellers from knowing what physical conditions obtained there that they conceived them to be desert, and, without exception, travellers had only the vaguest and most fantastic notions about the inhabitants of that zone. As late as the sixteenth century they were described, by Duarte Pacheco, as having 'the faces and teeth of dogs and tails like dogs'.[1] As for the location of their gold marts, where the

[1] *Esmeraldo de Situ Orbis, op. cit.*: I, 29 and Appendix No. III. The pre-dynastic Egyptians ornamented themselves with tails, as do some modern Negro tribes for ceremonial occasions, while the prognathous negro frequently admits of the description 'dog-like'.

exchange of salt for gold was carried out silently and often without any direct intercourse between the bargaining parties, nobody could say anything for certain. Pacheco could only give the distance they were from the coast, but not the direction. Al-Idrīsī only tells us that they were eight days' journey from Ghana—which is a lost city—again without indicating the direction. Cadamosto[1] is scarcely more helpful. However, when it is recalled that these 'frontier' negro tribes were frequently being raided by their northern Moorish neighbours and sold into captivity, it is small wonder that they should prefer to remain unseen when bartering gold, or that they should misrepresent the character of their mineral-producing hinterland—when caught.

Repulsed, but nothing daunted by the geographical obstacles to their overland advance, traders soon turned their attention to the navigation of the adjacent seas. In this the Arabs were to play but little part, for they had an ingrained horror of the Atlantic—'the Green Sea of Darkness'. Even Idrīsī dwelt at length upon the thick and perpetual darkness brooding over the Western Ocean. The doctors of the Koran held that a man mad enough to embark on its waters should be deprived of civil rights. Ibn Sa'īd went further, asserting that no one had ever done this for fear of the whirlpools which would most certainly destroy any adventurer.[2] As late as the generation immediately preceding that of Prince Henry the Navigator, another light of the Moslem world declared that the Western Ocean was

> boundless, so that ships dare not venture out of sight of land, for even if the sailors knew the direction of the winds, they would not know whither those winds would carry them. As there is no inhabited land beyond, they would run the risk of being lost in mist, fog or vapour.[3]

All the same at least one Arab, Ibn Fathima by name, ran

[1] *Fl.* c. 1450; *vide infra*, pp. 108–9.

[2] This, it will be remembered is precisely the fate which Dante assigns to Ulysses as a result of his mad flight beyond the Pillars of Hercules. *Vide Inferno*: XXVI, 124 *et seq*.

[3] Quoted by C. R. Beazley: *Prince Henry the Navigator*, pp. 14–15.

that risk and lived to tell the tale. Since the account of his voyage, which, paradoxically enough, is reported by Ibn Sa'īd, is circumstantial and not without its significance for the history of African exploration (for Cape Branco was rounded on the voyage), we will quote it at length.

Ibn Fathima relates how that, on one occasion, he found himself at Noul Lamtha near the shores of the Environing Sea and embarked upon a waiting ship. The vessel was eventually shipwrecked, going ashore on some shoals. The sailors had lost their way, and not knowing where they were, they abandoned the ship and went in a row-boat to reconnoitre. At times the boat made its way through patches of marine plants; at others it was raised out of the water by the oars (*sic*). Having arrived in the middle of the Gulf [of Arguim], the sailors were astonished at the great quantity of tunny which were living in those parts. They had not yet sighted land when they found that their provisions were exhausted. On reaching the White mountain some Berbers of the tribe of Godala made signs to them not to approach the mountain; the sailors did not understand the motive of this action. Nevertheless, they turned northwards and succeeded in rounding the mountain. Then a man appeared who could speak two languages, Arabic and Berber . . . They recounted what had happened. . . . Then they bought some horses and went to the chief town of the Godala tribe. This is the town of Tegazza, 11° of longitude and 20° north.

Following Ibn Fathima, the country occupied by the Godala, although similar to other parts of the Sahara and sandy countries in general, is very suitable for sugar-cane. Indeed the country is watered by five rivers which descend from the White mountain. Ptolemy has made mention of these southern rivers.[1]

The grounds for identifying the 'White mountain' with Cape Branco (or Blanco) are these. There is no other cape along the desert coast that in any way merits the appellation of 'mountain'.[2] Further, the colour of the cliffs is distinctly

[1] *Vide* J. T. Reinaud: *La Géographie d'Aboulfeda*, Vol. II, pp. 215–16.

[2] This is not invalidated by the statement referring to the five rivers. Ibn Sa'īd is probably giving his own theory on the Ptolemaic river system emanating from the Mountains of the Moon. *Vide Africa Pilot:* Part I, p. 289 (1920 Edition.).

white and shoals are a feature of the immediate vicinity.[1] Again, the Gulf of Arguim to the south-east is a notable tunny ground, the tunny trade having supplied a livelihood for a colony of Moors there from time immemorial.[2] The reference to the wealth of marine plant-life strengthens this identification still further, for the sea in this part of the African coast is covered with aquatic plants to such an extent that it is hardly an exaggeration for Ibn Fathima to speak of the boat's progress being impeded. It was on this account that the Portuguese later gave the name of 'Sargaço'[3] to this coast.

Fathima's achievement does not appear to have been followed up. His contemporaries, we may take it, were frightened by the perils, rather than stimulated by the prospects of reward. Although he does not point the moral, it is more than probable—in view of the above-quoted warning—that Ibn Saʿīd reported the voyage with the object of deterring any emulators.

Plainly, the Moslem kingdoms had their hands too full with the pursuit of trade and converts to have much time or inclination for solving mysteries that might, after all, prove to be mere will-o'-the-wisps. In consequence, we must look elsewhere than to the Arabs for the urge to explore.

This was supplied to the maritime peoples of Europe about this time by a variety of circumstances, among which the following are prominent. Firstly, the fall of Acre to the Saracens in 1291. This meant the final destruction of the hopes and schemes centring in a Latin land-dominion in the Levant. New outlets, in consequence, had to be found for the spirit of adventure and *Wanderlust* engendered by the Crusades. Secondly, the quest for the 'River of Gold', the illusory *El Dorado* of the Middle Ages. The popularity of this subject with the map-makers of the fourteenth and

[1] *Africa Pilot:* Part I, pp. 289–90. 1920.

[2] *Vide* A. Gruvel and A. Bouyat: *Les pêcheries de l'Afrique Occidentale française*, p. 115 *et seq.*

[3] Signifying 'gulf-weed' or 'sea-weed'.

fifteenth centuries is plain testimony to the importance which it assumed.[1] Thus the Norman Bethencourt's professed object in exploring lands to the south of Cape Bojador was, 'with the help of God, . . . to open the road to the River of Gold . . .' which, if he succeeded in doing, would have been 'greatly to the honour and profit of the kingdom of France'.[2] Thirdly, the growing menace of Islam to the peace and prosperity of Christendom. This was making men more and more desirous of finding the realms of the Christian king of Ethiopia—Prester John, who was also a popular subject with cartographers. With his aid it was hoped to frustrate the designs of the Moslems. Zurara lays it down as one of the avowed purposes of Prince Henry's maritime enterprise 'to seek to know if there were in those parts any Christian Princes, in whom the charity and love of Christ were so ingrained that they would aid him against those enemies of the faith'.[3]

What evidence is there of European enterprise along the West African coast in the pre-Portuguese period? The obscure thirteenth-century voyages of the Vivaldi brothers need not detain us long, for the information is fragmentary and, in part, unreliable. These men, it is said, set out in 1291 in two ships to 'go by sea to[wards] the parts of India and bring back useful things for trade'. All we know definitely is that they succeeded in passing a place called Gozora.[4] Cadamosto, the Italian sea-captain employed by Prince Henry the Navigator, writing more than 150 years after their time, affirms that they reached the 'Sea of Ghinoia'. Here one vessel was stranded, but the other sailed on to a city of Ethiopia called Menam, where lived Christian subjects of Prester John. None of the members of the expedition ever returned, but Usodimare who met Cadamosto on his first

[1] e.g. Catalan Atlas and Borgian World Map, *vide* Plate No. VIII.

[2] *Conquest of the Canaries* (Hakluyt Society): p. 106. *Vide infra*, p. 113.

[3] *The Chronicle of the Discovery and Conquest of Guinea*: Vol. I, p. 28 (Hakluyt Society).

[4] i.e. 'Caput finis Gozola' of the Pizzigani chart of 1367, i.e. Cape Non.

VIII

AFRICA—ACCORDING TO THE BORGIAN WORLD MAP (*ante* 1450)
(*Museum of Cardinal Stefano Borgia, Velletri*)

voyage declares that he spoke to the last surviving descendant![1] It is evident that very little weight can be attached to evidence of this sort. Beazley says of it that it 'looks like an attempt [by Usodimare] to amuse his creditors with interesting adventures'.[2] We may pass over the voyage of Jaime Ferrer in 1346 to the 'River of Gold' for no record of the expedition survives beyond the notice of its departure and purpose.[3] We may also dismiss the hypothetical French voyages of the same century, for the fabrication of the evidence for these has been proved once and for all by La Roncière.[4]

We come, then, to the anonymous fourteenth-century Spanish Franciscan,[5] about whose work *Libro del conoscimiento de todos los regnos y tierras* (*c.* 1350) opinions have differed greatly. Major, who knew of the work only from allusions to it in Bethencourt's *Conquest of the Canaries*, called it 'a confused compilation of the geographical traditions of the time and a réchauffé of Edrisi',[6] while Beazley holds it to be largely a record of an imaginary journey.[7] La Roncière says very much the same thing.[8] Markham, on the other hand, declares in his Introduction to the Hakluyt Society's edition of the work that 'there was nothing within his reach, from which the Franciscan could have pirated his narrative'.[9] In order not to misrepresent the claims of the Friar in the field of African exploration, we will cite his narrative in some

[1] *Le Navigazioni atlantiche di Alvise da Cá da Mosto* (edited by Rinaldo Caddeo), p. 153.

[2] *The Chronicle of the Discovery and Conquest of Guinea:* Introduction to Vol. II.

[3] *Vide* Plate No. VII.

[4] *La découverte de l'Afrique au moyen âge:* Vol. II, p. 10 *et seq.*

[5] All that is known of him is what he himself records in the opening paragraph of the work. 'I was born in the kingdom of Castille, in the reign of the very noble King Don Fernando . . . when the era . . . of Christ was 1304 years'.

[6] *Conquest of the Canaries:* Introduction.

[7] *The Dawn of Modern Geography:* Vol. III, p. 416, note.

[8] *op. cit.*, Vol. I, p. 118.

[9] *Knowledge of all Kingdoms and Lands* . . . (Hakluyt Society): p. x.

detail. He has just returned to Gazula (the Vivaldi's Gozora) from an extensive tour of Northern Africa and the Canaries when he finds some Moors who have

> armed a galley to go to the Rio del Oro, for there they make great profits. I went with them on account of something they said to me, so we left Gazula in that galley. Always keeping along the shores of the Western Sea we passed Cape Na (Non) and San Bin and Buyder (Bojador), all an uninhabited coast, and came to the Rio del Oro, already mentioned, which is a branch of the Nilus rising in the lofty mountains of the Antarctic Pole,[1] where, it is said, is the terrestrial paradise. . . . We went on, after leaving this river, for a very long distance always keeping in sight of the coast, leaving behind us the Islas Perdidas and came to an island inhabited by many people. They call this island Gropis.[2] It is a well-supplied island but the people are idolaters. They took us before their king and wondered much at us and our language and customs. The merchants who armed the galley made much profit. . . . We left the island of Gropis and shaped an eastern course towards the Southern Sea until we found another island called Quible.[3] This island . . . is in the South Sea and is peopled by Negroes. We left it on the right hand, keeping along the coast and came in sight of a very high mountain called Alboch.[4] We went there and found it inhabited by many people. There rises from it a river [5] which becomes very large and flows through a fine country. Here the galley turned back and I remained for some time. Afterwards I departed from Alboch with some people and went to another mountain called Lirry. A river rises from it called Enalco. I left this kingdom . . . and went to another called Gotonie consisting of some very high mountains. They say there are no others so high in the world. They are called the Mountains of the Moon.[6]

The Friar continues his narrative by telling of his travels in the land of Prester John, Ethiopia, Mesopotamia and the Far East.

[1] i.e. Southern Hemisphere.
[2] One of the islands in the Bissagos archipelago? Orango Island?
[3] Sherbro' Island? [4] Sierra Leone? [5] River Niger?
[6] *Knowledge of All Kingdoms . . .:* p. 34.

A cursory reading of the *Conoscimiento* would lead us to believe that the Franciscan was nothing more than a second Mandeville—a wholesale and unprincipled plagiarist. Major, Beazley and others, realizing that no man could possibly have accomplished all that the writer professes to have done, have been led to discredit the entire work. But a more careful examination of the narrative shows that such a judgment fails to do justice to all the facts. In the first place, the *Conoscimiento* is not a mere recapitulation of existing travel books, for it is impossible to find parallels for many of the descriptions in contemporary or earlier literature. Secondly, the narrative bears the stamp of credibility which is so often lacking in works like *The Travels of Sir John Mandeville*. There is, as in Ibn Batūta's travels, only the most economical use of the miraculous : thus, the Friar does not people the interior of the continent with *Sciapodae* or *Monoculi*—a feature all the more commendable in view of the appeal which the relation of such wonders made to the reading public of his day. Thirdly, the work is characterized by a *bonâ fide* mode of expression which seems to belie the charge of fabrication, witness such natural touches as 'I went with them on account of something they said to me', and 'They took us before their king and wondered much at us and our . . . customs'; again, 'The merchants who armed the galley made much profit.'[1] In addition, it must be admitted that his description of the Guinea coast has the important quality of verisimilitude. The information advanced, while it is slight, points to a knowledge only to be derived at or near the locality in question.

To reject the claims of the narrative is to create as many difficulties as it solves. For how could the knowledge have been transmitted to the Friar ? To say, as La Roncière does, that the Franciscan collected his material from Arabs, is to be at variance with the general tenor of Arab literature on West Africa. This is conspicuous for its ignorance of the

[1] The extremely plausible fiction literature of the *Robinson Crusoe* type finds no counterpart in the Middle Ages.

southerly part described by the author. Moreover, he reveals a strange lack of knowledge of the interior of the continent, a particularly curious fact if his informants were Arabs. Markham suggests, alternatively, that the 'knowledge may be due to contact with members of Genoese and Venetian caravans which penetrated far into the interior from Cairo'.[1] This lays itself open to a similar criticism, namely, that no accurate notion of the lands to the south of the Senegal-Niger system, or of the sea coast beyond, comes down to us from the accounts of medieval travellers and merchants in Africa. Neither Anselme d'Isalguier of Toulouse, who lived at Gao from 1405 to 1413, nor Antonio Malfante, the Genoese who wrote a description of the basin of the Niger in 1447, possessed any knowledge of the lands to the south. Malfante records that he had endeavoured to ascertain the place of origin of the gold, but he was always unsuccessful.[2]

Setting these considerations aside, there can scarcely be any quarrel as to the extent of the Friar's knowledge of West Africa. Briefly stated it is this:

(1) A fairly intimate knowledge of the western sea-board to Cape Bojador.

(2) A less certain—but still indisputable—knowledge of the coast beyond, possibly to some such island as Sherbro', and the region of Sierra Leone.

(3) A shadowy—perhaps wholly speculative—idea of the Guinea Coast, amounting to little more than acquaintance with its easterly trend.

(4) An almost complete ignorance of the interior of West Africa; thus the Senegal-Niger country is passed over with the barest mention.

A most illuminating commentary on the *Conoscimiento* is found, about fifty years later, in the work of Bontier and Verrier, historians to Bethencourt's Norman expedition to the Canaries in 1402 and following years. In this—*The History of the Conquest of the Canaries*—the Franciscan's state-

[1] *Knowledge of all Kingdoms . . .:* p. xii.

[2] For texts, *vide* Charles de la Roncière, *op. cit.* Vol. III, pp. 6 *et seq.*

IX

THE ESTE WORLD MAP (C. 1450)

(*Biblioteca Estense, Modena*)

ments, so far from being called into question, are accepted in good faith and made to serve as the *raison d'être* for an examination of 'the country from Cape Cantin which is half-way between the Canaries and Spain to Cape Bojador, a promontory to right of the Canaries extending on the other side to the Rio d'Ouro'.[1]

Thus it is apparent that the *Conoscimiento* had come to be regarded, within the space of a single lifetime, as a recognized text-book of African geography. From this *History*, too, comes clear testimony to the feasibility of the Friar's enterprise, for the writers speak of 'those who have explored those parts'.[2] It can be argued, of course, that Bethencourt's intention to 'open the road to the River of Gold'[3] is strong evidence against an antecedent discovery of the Senegal River. It is true that this statement militates against the idea of a fourteenth-century *French* discovery of the Senegal, but, on the other hand, it is no evidence against an expedition of another nation having succeeded in attaining the same goal.

A further proof of the credence attached to the *Conoscimiento* is forthcoming in the Este World Map of *c.* 1450, for there can be little doubt that the delineation of the West coast of Africa south of Cape Verde is there based upon data supplied by the Franciscan's narrative.[4]

Summing it up, it would appear that the documentary evidence for such a voyage, admittedly slight, is stronger than the cumulative evidence which can be advanced against it. It is perhaps not even too fanciful to suppose that the Friar made the expedition in company with Jaime Ferrer, whose goal was that of the merchants mentioned in the narrative—namely, the River of Gold. In any case, the seafaring activities of the Moors along this coast are so little known that there is no room for a dogmatic rejection of the entire narrative.

[1] p. 95 (Hakluyt Society). [2] *Ibid.*: p. 96. [3] *loc. cit.*

[4] *Vide* Plate No. IX and Memoir, by G. H. T. Kimble, accompanying the Royal Geographical Society's reproduction of *The Catalan World Map of the R. Biblioteca Estense at Modena.*

Turning now to consider the medieval knowledge of West Africa as disclosed by the *mappaemundi*, we find difficulty, once again, in making any useful generalizations. It is possible, however, to be dogmatic at least on two points. In the first place, Cape Non ceased to be 'Caput finis Africa' about the middle of the fourteenth century. Secondly, the ocean coast as far as Cape Bojador (more correctly as far as the cove on its southern side) was known and mapped from the time of the Pizzigani Portolan chart (1367).

Of more extensive knowledge we cannot be so certain, though various claims have been put forward. Dr. Hamy, for instance, in a memoir on the Dulcerto Portolan chart of 1339 declares that it shows the Majorcans 'possessed relatively exact knowledge on the ocean coasts . . . to the south of the Bay of Arguim which the Portuguese did not reach for 107 years afterwards'. The arguments for this conclusion are of the flimsiest. The African coast, south of Cape Non, is indicated only by a line bearing south-south-east to the extremity of the map. Only very general information accompanies this 'shoreline'. Thus: 'the whole of this coast is uninhabited except by fishermen.' The occurrence near the southern extremity of the map of the word 'felle', which Hamy identifies with Tafouelli, a region of the Senegal basin, has little, if any, significance, since we find that the Pizzigani chart has the kindred legend 'fele Ganuya' *north* of 'Caput finis Africa'.[1] Nor is Dulcerto alone in this delineation, for the Catalan Atlas of 1375,[2] the Viladestes map of 1413, the Valseccha map of 1439 and others, adopt a similar convention for their African coasts beyond Cape Bojador. The explanation is simple. Land exploration had, for a long time, outpaced maritime exploration, and so, as we shall see later, the map-makers of the fourteenth and early fifteenth centuries knew much more about the Sahara and the Sudan than they did about the Atlantic coast. Accordingly, in their efforts to indicate the great trans-Saharan caravan routes as accurately as possible, they found themselves

[1] i.e. Cape Bojador. [2] *Vide* Plate No. VII.

X. AFRICA—ACCORDING TO THE LAURENTIAN WORLD MAP (C. 1351)
(*Biblioteca Laurenziana, Florence*)

speculating on the trend of the coast south of Cape Bojador. By a fairly general consensus of opinion it was given a south-south-east direction.

The most extravagant claims put forward concern the Laurentian World Map of *c.* 1351.[1] For us its immediate source of interest lies in the delineation of the southern shores of Africa. Alexander von Humboldt, Baron Nordenskiold, La Roncière and Von Wieser all aver that the map exhibits in a greater or a less degree a foreknowledge of the continent's true shape. Santarem, on the other hand, finds in it 'proof of the indubitable priority'[2] of Portuguese discovery along the African coast.

The following facts emerge from a study of the African portion. Firstly, all details such as legends, rivers and towns stop short of the parallel of 'Mõs Lune'.[3] Secondly, this parallel coincides with a definitely marked line of juncture, south of which it is possible to detect, despite the mellowing influence of time, a difference of shade and brushwork. Now, as is well known, the territory of the normal Portolan chart of the late fourteenth and early fifteenth centuries did not extend farther south than Sierra Leone. The opinion that this line of discontinuity is more than a figment of the imagination is strengthened on a scrutiny of the northern portion of the map. Here, the author, instead of essaying to solve the problem of the Arctic, terminates his work in the latitude of central Sweden, i.e. just where his knowledge of the North might be expected to cease. The region north of this latitude is left entirely blank—a fact suggesting that the African side of the map may have terminated originally on its southern

[1] *Vide* Plate No X. This world map, properly called the *Portolano Laurenziano-Gaddiano*, forms part of an atlas, consisting of 8 sheets, housed in the Laurentian Library in Florence. For a fuller account of it, *vide The Laurentian World Map . . .*, by G. H. T. Kimble in *Imago Mundi*, Vol. I, 1935.

[2] *Recherches sur la priorité de la découverte des pays situées sur la côte occidentale d'Afrique :* p. 89.

[3] i.e. Montes Lunae which the cartographer has located in the hinterland of Sierra Leone.

front, in the neighbourhood of Sierra Leone. It is only fair to point out, however, that the break is not continued eastwards across southern Asia. In the third place, the map carries a second outline of the shape of the continent. This gives an entirely different conception of Africa, one, indeed, that is more nearly in agreement with medieval map-practice, as exemplified in the Sanuto and Mauro planispheres. As this outline seems to have been superimposed upon the already existing ocean, it may be regarded as having been drawn after the world map first appeared. Which delineation was first in point of time can hardly be decided from inspection.

How are we to regard these interpolations? Are they the work of an editor, or editors? The merchants of Genoa, it goes without saying, would evince the greatest interest in the matter of the southern extension of the continent, particularly as their land intercourse with the East was being interrupted by the Turks. Consequently, the portrayal of south Africa may conceivably be an attempt on their part to solve the age-old question of 'Africa nondum cognita'. Alternatively, it may be that the editing, or part of it, was done at the Portuguese Court in the following century, after Prince Pedro, brother to Prince Henry the Navigator, had returned from Italy. On that occasion he brought home a number of maps by which 'he was much helped and furthered in his discoveries'.[1] It has even been suggested that the Laurentian World Map was among these.[2] If this is so, the 'modern' rendering had been given to it in Portugal, at the time when the true shape of the continent was being disclosed. In deciding about the outline map we are helped by an early fifteenth-century map—the world map of Albertin de Virga, dated 1415.[3] A comparison of the two maps seems to indicate that the Laurentian editor, or author, was conversant with the later work and that he was sketching his outline either from memory or directly from this map. The former is

[1] Antonio Galvão: *Discoveries of the World* (Hakluyt Society), p. 66.
[2] *Vide* J. R. Spears: *Master Mariners*, p. 63.
[3] *Vide* Plate No. XI.

the more probable since the correspondence between the two outlines is not very close, but any suggestion that the similarity is adventitious is seen to be unlikely when we realize that no other medieval world maps introduce the Terrestrial Paradise under exactly the same guise.

There is no need now to go into the arguments that have been advanced in favour of a later date for the map.[1] Investigation shows that there is ample reason for retaining the date of *c.* 1351, and for regarding the two interpretations of the African outline as later emendations, the first being made soon after the appearance of the de Virga map in 1415 and the second at considerably later date, presumably when the Portuguese were well on their way to the Cape.[2] Apart from such an explanation the Laurentian World Map would remain an enigma, for neither the fourteenth nor early fifteenth century holds any record of achievement which can account for such a revolutionary change in geographical ideas as this map exhibits in its 'modern' delineation of the African continent.

Almost equally 'modern' in outline, however, is the Africa of Fra Mauro's famous Planisphere of 1459,[3] for, putting aside for the moment questions of interpretation, it is impossible not to be struck by the illusion of accuracy which the general shape of the continent produces. More than one student of cartography has shared Alexander von Humboldt's conviction that the southernmost point, called 'Cavo de Diab,' is none other than the Cape of Good Hope made known by some daring expedition similar to that which Mauro himself

[1] e.g. by F. von Wieser in *Die Weltkarte des Albertin de Virga aus dem Anfange des XV. Jahrhunderts.*

[2] *Vide The Laurentian World Map . . . : loc. cit.*, pp. 29–33.

[3] Little is known of the life of Mauro beyond the fact that he was a monk of the Order of the Camaldules at S. Michele di Murano, near Venice, that he attained a considerable distinction in mathematics and physics, and that he was an 'incomparable cosmographer'. It is probable that the Planisphere was constructed at the demand and cost of the Venetian Government.—*Vide* P. Zurla: *Il Mappamondo di Fra Mauro Camaldolese, passim.*

speaks of as having taken place in 1420.[1] But an analysis of the features of the hinterland makes it clear that they do not refer to *south* Africa, for the Mareb and Tagas rivers with their affluents, Mana, Lare and Abavi, can be none other than the Abyssinian rivers, Mareb, Takkazye, Menna, Tellare and Abbai; while 'flumen Xebi' and 'flumen Avasi' are traced with such extraordinary fidelity that they can be readily identified with the Ghibie and Hawash rivers of Southern Abyssinia—a region not thoroughly explored until recent times. The fact that these rivers are the southernmost feature of the map—they are placed in the same latitude as 'Cavo de Diab'—makes it almost impossible to believe that Mauro knew anything of Africa south of the Equator. How, then, are we to interpret Diab, the hinterland of the Cape, described by the map-maker as 'a very fertile region which was recently conquered by the great King of Abyssinia, *c.* 1430'? The only region with which it is at all comparable, within the limits imposed by Mauro's knowledge of the interior, is the Somali peninsula, and it is significant that in a contemporary document in the library of S. Michele at Murano we are told that 'Diab is a great province, in parts of which there is an abundance of every good thing, its principal town being called Mogadis'[2] which can be none other than Magadoxo of the Somali coast. 'Cavo de Diab,' in that case, becomes Cape Guardafui.[3] Consequently, the degree of verisimilitude

[1] 'About the year 1420, an Indian vessel, or junk, which was on her passage across the Indian Ocean to the Islands of Men and Women, was caught by a storm and carried for 40 days 2,000 miles beyond Cavo de Diab to the west and southwest, and when the stress of the weather had subsided, was 70 days in returning to the Cape.'

[2] *Vide* Charles de la Roncière: *op. cit.*, Vol. II, p. 133.

[3] It will doubtless be objected that Diab is represented as an island and not as a peninsula. In extenuation of this it may be urged that Mauro's purpose, throughout, seems to have been to harmonize conflicting theories, except, of course, where they were mutually exclusive. Here, we may take it, he attempts to incorporate the Arab idea of a channel connecting the Indian and the Atlantic Oceans (*vide ante*, chap. iii, p. 55) with his information based on 'the drawings of travellers who have been there'. As these all appear to have been *land* travellers, it is highly probable that

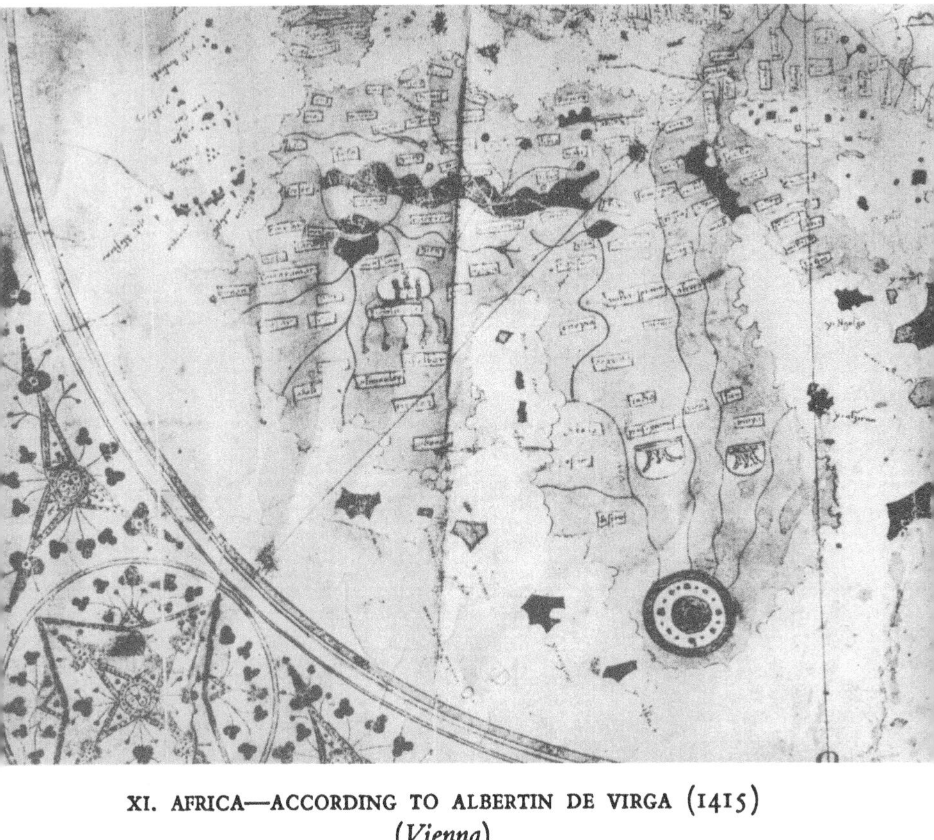

XI. AFRICA—ACCORDING TO ALBERTIN DE VIRGA (1415)
(*Vienna*)

possessed by the southern portion of African coastline is due largely to the exigencies of the map's circular form.[1]

In conclusion, then, it may be averred that, despite the dearth of confirmatory map evidence, and the firm Portuguese belief to the contrary, the medieval knowledge of Africa, especially of the coasts, extended to well within tropical latitudes. Measured solely by the criterion of mileage these early explorers cannot be said to have added very much new territory to the map of the world. Certainly their voyages do not compare with those of the following age, but without their pioneers the achievements of Columbus, da Gama and Magellan would not have been possible. Sailing 'by God and by guess', with no pilot-books and few charts, their only instruments a compass and an astrolabe, these unknown mariners were responsible, perhaps more than anybody, for the passing of the Middle Ages and the coming of modern times.

they had nothing to contribute to the meagre fund of coastal knowledge of East Africa, and, consequently, that Mauro was left to his own devices.

[1] *Vide* Plate No. XII.

CHAPTER VI

EXPANDING HORIZONS—ASIA

IN the literature of the early Middle Ages nowhere is the poverty of geographical knowledge more apparent than in regard to the Far East. Whereas in the writings of the Ptolemaic era we are furnished with relatively precise information on such topics as the mountains of Central Asia, the characteristics of the Chinese people and their silk trade with the West, in the centuries that follow the text-book frontiers of Asia are withdrawn to the Ganges, and in everything but name China is unknown.

Asia [writes Orosius in the fifth century] begins in the East where lies the mouth of the River Ganges, facing the Eastern Ocean. On its left is situated the promontory of Caligardamia [1] which is hard by the island of Taprobane.[2] Here the sea begins to be known as the Indian Ocean. On the right [3] are the Imaus Mountains [4] where the Caucasus takes to itself the headland of Samara.[5] Adjoining this to the north is the mouth of the River Octogona.[6] From this point the sea becomes known as the Seric Ocean. Within this area lies India; on the west it is bounded by the River Indus which flows into the Red Sea; on the north, by Mount Caucasus; the rest of India is circumscribed by the Eastern and Indian Oceans. This country has 44 different peoples (if we include the island of Taprobane which has 10 states and the many other habitable islands). From the River Indus which is in the east to the River Tigris which is towards the west, there are the following districts, all of them mountainous and rugged: Aracosia,[7] Parthia, Assyria, Persia and Media. On the north these are bounded by the Caucasus Mountains; on the south, by the

[1] Cape Comorin? [2] Ceylon. [3] i.e. to the north.
[4] Pamir Mountains. [5] Sarmatia? [6] Oxus River? [7] Afghanistan.

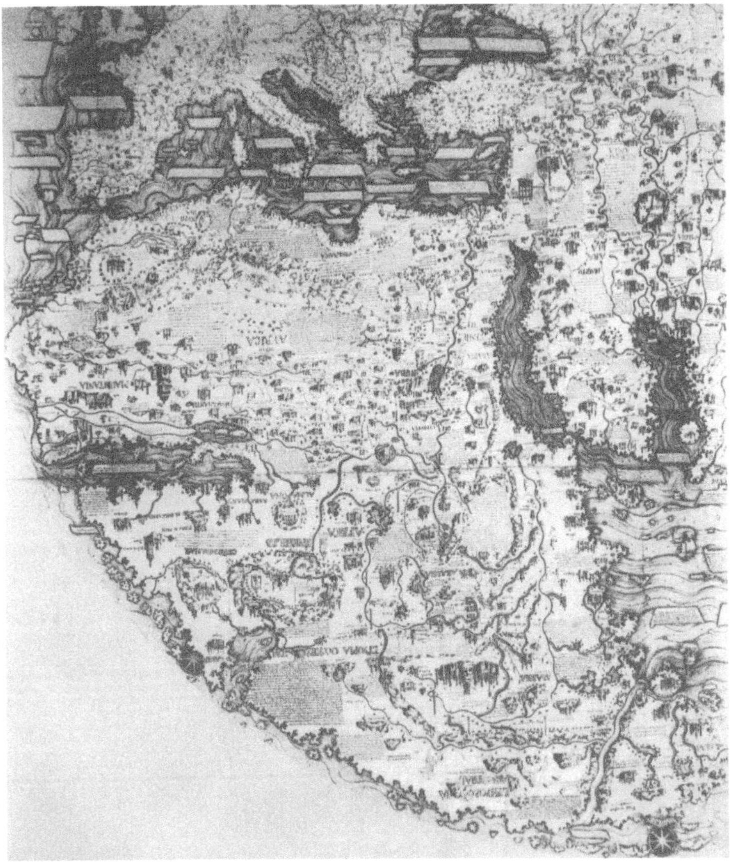

XII

AFRICA—ACCORDING TO FRA MAURO (1459)

(Biblioteca Marciana, Venice)

Red Sea and the Persian Gulf. The chief rivers flowing through the midst of these countries are the Hydaspes [1] and the Arbis.[2] Here there are 32 peoples. (The country which the Sacred Scriptures call Media is now usually called Parthia). Between the River Tigris and the River Euphrates lies Mesopotamia, a country beginning in the north midway between the Taurus and Caucasus Mountains ; in the south it is succeeded by Babylonia, then by Chaldea and finally by Arabia. This stretches eastwards in a narrow strip of land between the Persian and Arabian Gulfs. In these countries there are 28 different peoples.[3]

Set beside Ptolemy's *Geography*, Pliny's *Natural History* or even Ammianus Marcellinus' *History*,[4] statements of this kind —and they are quite representative of their century—invite unfavourable comparisons.[5] The loss of knowledge in the interval comes as something of a surprise when we recollect that the Byzantine Empire was still in touch with Turkish tribes of Central Asia in the fifth century, and that under Justinian the overland traffic in silk was almost as flourishing then as it had been in the first century.[6] The explanation lies, it would seem, with the changed mentality of the times. Christendom was preoccupied with strengthening the faith of its converts ; it had little time for pursuits ostensibly devoid of religious significance.

Shortly after the smuggling of the first silk moths into Europe had been accomplished in the middle of the sixth century, the trade with China—always hazardous and slow—began to dwindle and intercourse between East and West

[1] River Jhelum, tributary of the Indus River.

[2] i.e. River Arabis (= Purali) of Alexander's expedition.

[3] *Historia adversos paganos . . . :* I, 2.

[4] cf., for instance, the descriptions of the land of Seres in *Geography*, VI, 16 ; *Natural History*, VI, 20, and *History*, XXIII, 6.

[5] The *Cotton* map (Cotton MSS. Tib. B.v. fol. 56v) gives us some idea of Orosius' world-picture, for it is based very largely on his geographical matter and may even have been taken from the now-lost Orosian *mappamundi*. *Vide* C. R. Beazley : *The Dawn of Modern Geography*, Vol. II, p. 562.

[6] *Vide* G. F. Hudson : *Europe and China* . . ., p. 109 *et seq*., also Cosmas Indicopleustes : *Christian Topography* (Hakluyt Society), p. 47 *et seq*.

became, consequently, exceedingly uncommon. Between the seventh and twelfth centuries, with a solitary exception, there is no Latin or Greek work which reveals a living knowledge of China.[1] The exception is provided by Theophylactus Simocatta, an Egyptian Greek who, writing *c.* A.D. 628, speaks of ' the land of Taugas ' bordering upon that of the nomadic Turks. The people of Taugas,[2] he tells us, are one of the greatest nations of the world in power and population ; they are idolaters, but possess just laws ; and they produce silk and carry on a great commerce. He speaks further of their land being divided into two parts by a mighty river,[3] and of a recent war between rival states on either side of the river ending in the overthrow of one of them.[4] But Simocatta's *History*[5] enjoyed little or no vogue and so the textbook geography of the Far East remained historical and out of date. Bede in the eighth century, Dicuil in the ninth, Gerbert in the tenth, Adam of Bremen in the eleventh and numerous other Christian writers in the twelfth century merely retell the Orosian tale.

Meantime the Arabs were beginning to busy themselves with missions and merchandise in India and the lands beyond. These activities soon produced a literature which, for size and merit, equalled the later narratives of the Polos and their compatriots. The precise date of the entrance of the first Arabs into China is not known, but it may reasonably be assumed that it was not later than the beginning of the eighth century, for by A.D. 758 their numbers at the port of Canton alone were such that they were strong enough to attack and

[1] On the other hand, there are several Chinese documents which suggest that there was never a complete break in contact between the Orient and the Occident even in the dark seventh and eighth centuries. *Vide* G. F. Hudson : *op. cit.*, p. 128 *et seq.*

[2] i.e. the Tamghaj—name for China—of the Persians and Arabs.

[3] Yangtse-Kiang ?

[4] This, no doubt, refers to the struggle between the Sui and the Ch'en dynasties, which concluded in the victory of the former and the unification of China under their sway in A.D. 588.

[5] The Chinese passages are found in Book VII, 7 *et seq.*

pillage the city. In the same century Arabs also began to trade with northern China by an overland route, as well as by the sea route. There was, however, a considerable time-lag between this early prospecting and the systematic description of the new lands, for the earliest account extant is a compilation dating from the middle of the ninth and the beginning of the tenth centuries. The first part of it consists of a collection of notes about India and China,[1] based upon personal experience in the case of India and upon hearsay in the case of China. There is practically no geographical material concerning the latter country in this part of the work beyond the mention of Tibet as one of China's western neighbours and the presence of some islands—called Sila—to the east of the country.[2] The second part of the compilation was written by a certain Abu Saʿīd and is notable, *inter alia*, for its acquaintance with Khamdan,[3] the tenth-century capital of China, and with the overland route between Sogdiana and China proper. According to Abu Saʿīd's account, the frontier between these two countries was a two-months' journey distant over a waterless terrain. The difficulty of crossing this desert, so his informants maintained, had alone prevented Moslem armies in Khorasan from essaying to invade China. A friend of the author's had told him, however, that he had seen a man at Khanfu[4] with a bagful of musk on his back, who, he found, had come all the way from Samarkand on foot.[5]

[1] *Vide* H. Yule-H. Cordier: *Cathay and the Way Thither* (Hakluyt Society), Vol. I, p. 241 *et seq.*

[2] Idrīsī also speaks of these islands which seem to be identifiable with the Japanese archipelago, although Cordier maintains that Korea offers a more likely alternative. *Vide op. cit.*, Vol. I, p. 131 *note* and p. 257 *note.*

[3] i.e. Ch'ang-ngan—now represented by Si-ngan-fu in Shensi—which was the capital of several Chinese dynasties between the twelfth century B.C. and the tenth century A.D.

[4] Hangchow ? Canton ?

[5] *Vide* J. T. Reinaud: *Relation des voyages faits par les Arabes et les Persans dans l'Inde et à la Chine dans le IXe siècle de l'ère chrétienne* (Text and translation), Vol. I, p. 114.

More methodical, if at the same time more ambiguous, is the almost contemporary account of China by Ibn Khurdādhbih. In his *Book of Routes and Provinces* [1] he tells us much that is self-evidently true of the life and commerce of the country. Thus:

[at] Al-Wakin which is the first port of China [2] . . . you find excellent Chinese iron, porcelain and rice. You can go from Al-Wakin, which is a great port, to Khanfu in four days by sea or in twenty days by land. Khanfu produces all sorts of fruits and vegetables, wheat, barley, rice and sugar cane. . . . In all the ports of China you find a great navigable river affected by the tide . . .[3]

The length of coast-line from one extremity of China to the other is given as two months' journey. The country can boast of

300 prosperous and famous cities. It is bounded by the sea, by Tibet and by the country of the Turk . . . What is beyond China is unknown. But in front of Kantu [4] rise high mountains. These are in the country of Sila which abounds in gold . . . The products exported [include] gum, aloes, porcelain, satin and cinnamon . . .[5]

Whether Khurdādhbih actually visited China is not clear, but there can be no doubt from these extracts that he was in possession of authentic information concerning that country. All the crops catalogued by him were characteristic of the medieval marts of the Far East.

With so much already known by the Arabs in the tenth century, we are encouraged, not unnaturally, to expect that subsequent writers will be able to add further details. But neither Al-Masʿūdī, Al-Idrīsī nor Abū'l Fidā succeeds in telling us much that is new. Masʿūdī claims to have visited Siam, Java, China and even Turkestan, but his narrative betrays little or no personal acquaintance with those lands.

[1] *Vide ante*, chap. iii. [2] Canton?

[3] *Vide* H. Yule-H. Cordier: *op. cit.*, Vol. I, pp. 135–6.

[4] Shantung Peninsula?

[5] H. Yule-H. Cordier: *op. cit.*, Vol. I, p. 137.

As for Idrīsī, his account is disjointed and not always easy to follow. Moreover, while professing to give exact distances between places, he generally underestimates them. His references to Chinese commodities of trade agree, substantially, with those given by Khurdādhbih. So, too, do those relating to the chief ports; on the other hand, some of his place-names are more reminiscent of Ptolemaic than of contemporary geography.[1] Abū'l Fidā is hardly more helpful. His knowledge of the Far East is, as he himself complains, very much restricted and is derived chiefly from books written long before his time. They are a strange mixture of obsolete statements of earlier Arab geographers and fragments of recent history. He tells us, for example, that

> China is bounded on the west by [desert] lands separating it from India; on the south by the sea; on the east by the eastern Atlantic; on the north by the lands of Gog and Magog and other regions respecting which we have no information . . . there being few travellers who arrive from those parts, such as might furnish us with intelligence [respecting those places] . . . we forbear to detail them . . . Khanfu is one of the gates (i.e. ports) of China and is situated on the river . . . Ibn (i.e. Abu) Saʿīd says it is mentioned in books and it is situated on the east of the river of Khamdan. Ibn Khurdādhbih says it is the greatest commercial port of China and abounds in fruit, vegetables, wheat, barley, rice and sugar-cane.[2]

In Ibn Batūta[3] we have a writer of a different calibre. A great traveller, as we have already seen, he recorded his observations on the 'celestial' Kingdoms of Cathay and Manzi,[4] and the romantic Middle East without undue re-

[1] e.g. Katighora (Cattigara), Asfiria (Aspithra ?), Taugha (Taugas).

[2] H. Yule-H. Cordier: *op. cit.*, Vol. I, p. 257.

[3] *Fl. c.* 1340.

[4] From *c.* A.D. 960–*c.* 1280 China was nominally ruled by the Sung Dynasty. Its effective government, however, was confined to southern China. Northern China, meantime, was for the most part under the sway of nomads from Manchuria. It was during this period that the names of Cathay and Manzi came into use to denote North and South China respectively. Persisting under the Mongol Empire, they were adopted by Occidentals. At times 'Cathay' was extended to cover China as a whole.

course to legendary embellishments. At the same time, his narrative is lamentably poor in geographical understanding; to look for systematic regional descriptions—even for itineraries which respect chronology—is to be disappointed. The travels were the thing, their description an afterthought. In as far as Batūta deserves the title of geographer, it is in spite of himself. His observations on China are typical of his style and method. It is a land, he tells us, 'of vast extent and abounding in produce, fruits, grain, gold and silver. In this respect there is no country in the world that can rival it.' The port of Zaiton[1] is, in his view,

> one of the largest in the world, or perhaps the very largest . . . It is formed by a large inlet of the sea which penetrates the land to the point where it unites with the great river.[2] In this city, as in all Chinese towns, a man will have a fruit garden and a field with his house in the middle of it . . . For this reason their towns are extensive. . . . Cathay is the best cultivated country in the world. There is not a spot in the whole extent of it that is not brought under cultivation (*sic*). The reason is that if any spot is left uncultivated its inhabitants or their neighbours are assessed for the land-tax due thereon. Fruit gardens, villages and fields extend along both banks of this river without interruption from the city of Khansa[3] to the city of Khan-Baliq[4] which is a space of 64 days' journey . . . I have never seen anything in the world like it, except a space of four days' journey between Anbar and 'Ana (in Iraq). We used to disembark every night and stay in the villages. . . .[5]

Although Ibn Batūta is obviously giving us some authentic information about this era it is not entirely free from error, for China boasts of no river flowing from north to south across the plains of Kiang-su and Shantung, but only a canal which no one who had ever sailed on it was likely to mistake for a river. So confused, indeed, are his ideas on inland

[1] Chuan-chow near Amoy, *vide infra*, p. 138.

[2] Hardly an appropriate designation for a stream only 50 miles long.

[3] Khinsai? [4] Peking.

[5] *Ibn Batūta's Travels*, H. A. R. Gibb's translation (*Broadway Travellers*): pp. 282 *et seq.*

China that he informs us in another place that there is a river 'called the Water of Life which rises in some mountains . . . near Khan-Baliq and flows through the centre of China for the space of six months' journey, until it finally reaches Sin as-Sin',[1] i.e. Canton. Again, he assures us that elephants are used as beasts of burden in Cathay—a fact which has led one scholar to conclude that Ibn Batūta never went to China at all.[2] It needs to be borne in mind, however, that Batūta does not lay claim to have visited all the places he himself describes. Supposing, as seems not improbable, that he only visited the coastlands of China, his misrepresentation of the hinterland becomes understandable, for it would be easy with his imperfect interpretation of Chinese reports (themselves not always reliable) to confuse the Grand Canal system linking up the Hwang-Ho and the Yang-tse Kiang with a north-south flowing river system. Be this as it may, there can be no doubt as to the genuineness of most of his travels. As we have already seen in connexion with Africa, Batūta's descriptions of that continent have every appearance of accuracy, and the same is certainly true of India and even of such inconsiderable localities as the Maldive Islands.[3]

The Arabs had reserved their best wine until last. After Batūta they can boast of few Oriental travellers or writers of repute. Significantly, this sudden termination of their activity synchronizes with the fall of the Mongol power and the rise of the Ming Dynasty which resumed the traditional anti-foreign policy of the Chinese and made it practically impossible for foreigners, Moslems and Christians alike, to enter the country.

Disappointing as this turn of events was to the aspirations of the West it did not come before Christian travellers had emulated the Moslems in traversing the entire length of Asia and in revealing to Western Europe the glories of India and

[1] *Ibid.*: p. 282.

[2] *Vide* G. Ferrand: *Relations de voyages et textes géographiques Arabes, persans et turks relatifs a l'Extrême-Orient du VIIe au XVIIIe siècles.* Vol. II, p. 433.

[3] *Vide* H. A. R. Gibb: *op. cit.*, pp. 13–14.

the Celestial Empire. Thanks largely to the conquests of Chinghiz Khan and his immediate successors, it became possible in the thirteenth century to travel across Asia from the banks of the Volga to the Yellow Sea without much more danger than normally attended the progress of the Christian pilgrim from Canterbury to Compostella. At first there was a certain excusable reluctance on the part of Europeans to use a route which passed through the territory of a people who at one point bade fair to plunge Christendom into a new barbarism. Later, when the fear of invasion had abated, the impression seems to have gained ground in Europe that these barbarians were ripe for conversion. Doubtless there was an ulterior motive here, for the greatest need of the Church at the time was to find a counterpoise to the Saracen power, and who could suit her purpose better than these fighting hordes from the steppes? Arguments, aimed at giving effect to such an enterprise, were not difficult to find. In the first place, it was discovered that the Mongols were quite tolerant of Christians residing in their territories. Then rumour had it that some of the Khans and their captains already professed the rudiments of the Christian faith. Thirdly, there were the stories of the Priest-King Prester John. In their earliest form, these located him in the 'Indies'.[1] Now India had always held a peculiar charm for medieval men, being associated in their minds with fantastic wealth, natural wonder and magic. It was said by many that the body of the miracle-working St. Thomas was buried there. Accordingly, the twelfth- and thirteenth-century stories about an Indian ruler who combined the piety of the Apostle with the wealth of a Croesus simply served to corroborate the romantic inventions which already fed the imagination of Europe and figured in the contemporary world-maps.[2] Such

[1] A vague term, for in the Middle Ages there were at least three Indias, viz. India Minor, India Major and India Tertia, i.e. the Sind, Hind and Zinj of the Arabs. The first two were located in Asia, the last in Africa (in Ethiopia).

[2] e.g. Hereford, Ebstorf and Este maps, *Vide* Plate No. IX.

was the trust put in these stories that it is no exaggeration to say that the quest for the realms of Prester John soon became, as it was to remain for more than a century, a major factor in promoting Asiatic enterprise. For this reason, it deserves our more detailed attention ; indeed, many a real character in the history of exploration has less claim to it than this half-legendary figure.

The earliest authentic reference to the Prester in Western literature occurs in the Chronicle of Otto, Bishop of Freisingen. Here it is stated that when he was at the papal court in 1145 he met the Bishop of Gabala [1] who related how

> not many years before, one John, king and priest (rex et sacerdos), who dwelt in the extreme Orient beyond Persia and Armenia, and was with his people a Christian, but a Nestorian, had made war against his brother kings the Medes and Persians, who were called Samiards [2] and captured Ecbatana their capital. After this victory Presbyter John—for so he was wont to be styled—advanced to fight for the Church at Jerusalem ; but when he arrived at the Tigris and found no means of transport for his army, he turned northwards, as he had heard that the river in that quarter was frozen over in winter time. After halting on its banks for some years (*sic*) in expectation of a frost, he was obliged to return home.[3]

It was further said that the Presbyter was descended from the race of the Three Wise Men, that he ruled over the same nations that they ruled and that he had such wealth that he used a sceptre of solid emerald.[4]

[1] i.e. Jibal in Syria. [2] Or Sanjards.

[3] *Ottonis Frisingensis Chronicon :* VII, 33. *Vide* Yule-Cordier : *op. cit.*, Vol. III, p. 16 note.

[4] The identity of this potentate has been the subject of much controversy. A. d'Avezac (*Recueil de voyages et de mémoires, publié par la Société de Géographie*, Paris, Vol. IV) and G. Oppert (*Der Presbyter Johannes in Sage und Geschichte*) have shown that he may have been Gur Khan, or Kor Khan, who held a great dominion in Central Asia, having as its centre Balasaghun, a town situated north of the Tian Shan Mountains. In 1141 the aid of this ruler was invoked by the Shah of Kharezm against Sanjar, the Seljuk lord of Persia who had expelled the Shah from his kingdom and killed his son. The Gur Khan replied by sending an army of Turks and others and defeated Sanjar near Samarkand. The chief obstacle to

Considerable as was the impression which this report made upon the leaders of the Western Church, it was far exceeded, twenty years later, by the circulation of a letter purporting to be addressed by Prester John to the reigning Byzantine Emperor Manuel I. It represented this monarch as ruling over the three Indias and described his powers and possessions in the most flamboyant terms. It began thus: 'John, Priest by the Almighty power of God and the strength of Our Lord Jesus Christ, King of Kings, Lord of Lords . . .' Continuing, the writer claimed that he was the greatest monarch under heaven, as well as a devout Christian. The letter dealt at length with the millennial glories of his territory:

> Honey flows in our land and milk everywhere abounds. In one region there no poison exists and no noisy frog croaks; no scorpions are there or serpents creeping in the grass . . . In one of the provinces flows a river called the Indus which, issuing from Paradise, extends its windings by various channels through all the province; and in it are found emeralds, sapphires, carbuncles, topazes . . . and many other precious stones.[1]

How great was the popularity and diffusion of this letter may be gauged, in some degree, from the fact that Zarncke, in his treatise on Prester John, gives a list of nearly one hundred manuscripts.[2] Incredible as it may seem to us, the genuineness

the acceptance of this identification is that there was no profession of Christianity on the part of Gur Khan: in fact he was a Buddhist. (But according to Sir Henry Yule there were constant confusions of rumour between these two religions as they existed in Central Asia; *vide Cathay and the Way Thither*: Vol. III, p. 22.) Bruun, on the other hand, thinks that he finds a warrior answering better to the indications in a Caucasian prince of the twelfth century, John Orbelian by name, who was a Christian. *Vide* Yule-Cordier: *The Book of Ser Marco Polo* (3rd edition), Vol. I, pp. 232 *et seq.*

[1] Quoted in full by Sir E. Dennison Ross in *Travel and Travellers in the Middle Ages* (edited by A. P. Newton), pp. 174 *et seq.*

[2] *Vide* F. Zarncke: *Der Priester Johannes* in the *Abhandlungen der phil.-hist. Classe der K. Sächsischen Gesell. d. Wissenschaften*, Vols. VII, VIII, 1879, 1883. Eight of these are in the British Museum, ten in Vienna, thirteen in Paris and fifteen in Munich.

of the document was never impugned by the great majority of people. At the same time, it did not produce any immediate response and when, in 1177, Pope Alexander III did dispatch his physician and friend, Master Philip, to find the Priest-King, it was not so much in answer to the letter as on account of reports received from Philip and others to the effect that Prester John was anxious to be instructed in the true Catholic faith. The letter borne by the doctor is preserved for us, in slightly varying forms, in the chronicles of several annalists, for example Matthew Paris [1] and Roger Hoveden.[2] In this the Pope recites how he had heard of the monarch's Christian profession, his pious desire to have instruction in the faith and to possess a church in Rome and an altar in Jerusalem. He found it impossible, on account of the length and difficulties of the way, to send any one 'a nostro latere', but he would send Philip to communicate instruction to him.[3] It is not stated that Philip had actually been to the Prester's court, but only that he had heard of his majesty's pious wishes from conversation with honourable persons in his kingdom, whom the physician had met 'in illis partibus'. The exact interpretation of this last phrase is a matter for doubt. Sir Henry Yule accepting Baronius' interpretation of the letter—Hoveden's version—that the 'King of the Indians, vulgarly called Pretejanni, reigning far and wide over Ethiopia' was the King of Abyssinia, is of opinion that the phrase may refer to Egypt or to Palestine, where Philip might well have met with Abyssinian pilgrims. The weakness of this contention, as Yule is quick to point out, is that the application of the title Prester John to the Abyssinian King was probably a good deal later. Further, in a letter reproduced by Matthew Paris

[1] *Chronica maiora*, edited by H. R. Luard (Rolls Series, No. 57), Vol. II, pp. 316–17.

[2] *Chronica*, edited by W. Stubbs (Rolls Series, No. 51), Vol. II, pp. 168–70.

[3] Licet autem grave nimis videatur et laboriosum existere ad praesentiam tuam inter tot labores et varia itineris locorum discrimina, et inter longas et ignotas oras, quemlibet a nostro latere destinare . . .' (Hoveden's version.)

—written by the Dominican Prior in Palestine in 1237—we read of one whose 'prelacy extended over India the Greater and the Kingdom of Prester John and other realms still nearer the rising sun'.[1]

Nothing more is heard about the Prester legend until 1221. In that year news came out of the East that a great Christian monarch was fighting the Moslems and sweeping away their power. The name ascribed to him was David, and some called him the son, others the grandson, of Prester John of India. Actually, this new conqueror was none other than the Tartar emperor, Chinghiz Khan. The delusion was speedily shattered, but it did little to damp the spirits of those intent on finding this vast monarchy. Popes and kings went so far as sending Franciscan missionaries to the Mongol hordes. While these messengers of the Gospel did not find their quarry, they made other discoveries which, from the standpoint of the extension of geographical boundaries, brought them ample compensation.

The credit for these discoveries rests, primarily, with two friars, John of Plano Carpini and William of Rubruck, in that they blazed the overland trail to Tartary. Although Carpini says next to nothing of Prester John, he speaks at length of the supposed piety of the Cathayans[2] who were said to have

> the Scriptures of the Old and New Testaments. They have also recorded in histories the lives of their forefathers: and they have hermits, and certain houses made after the manner of our churches. . . . They worship one God. They adore and reverence Christ Jesus our Lord, and believe the article of eternal life, but are not baptized. They do also honorably esteem and reverence our Scriptures . . .[3]

[1] *Vide Cathay and the Way Thither:* Vol. III, p. 18 note; also *The Travels of Sir John Mandeville: op. cit.*, p. 122.

[2] In Carpini's narrative the Kingdom of North China is given the name of Cathay for the first time.

[3] *The Voyage of Johannes de Plano Carpini:* chap. 9 (in Richard Hakluyt's *Navigations, Voyages and Discoveries*, 1598–1600 Edition).

The narrative then goes on to portray the habits and commerce of the people. William of Rubruck, on the other hand, speaks more explicitly and declares that the stories about 'King John' were grossly exaggerated by the Nestorian Christians whose king, so he maintains, the Prester had been. They report

ten times more of him than was true. For so the Nestorians which come out of those parts use to do. For they blaze abroad great rumours and reports upon just nothing . . . How-beit, when I travelled along by his territories, there was no man that knew anything of him, but only a few Nestorians.[1]

After Rubruck, came Marco Polo of Venice,[2] the greatest traveller of the Middle Ages, and perhaps of all time. But his mission to the Far East was diplomatic rather than religious and he makes little attempt to get to the bottom of the legend. In fact he only repeats, in the most guarded of terms, what he had been told about him.[3] This was by no means complimentary, for in Polo's day the honour of identification with Prester John, after hovering over one head and another, had finally rested upon Un-khan, the king of the Nestorian tribe of Kerait. By all accounts this man was a tyrant, exacting tribute where he could and reducing the Turks around him to slavery. As would seem to befit such a despot, he fell by the sword in battle against Chinghiz Khan.

After the time of Odoric of Pordenone,[4] who has a single sentence to the effect that 'the empire of Pretegoani' was fifty days' journey to the west of Cathay,[5] the priest-king ceases to have any pretensions to historical existence in Asia. When

[1] *The Journal of Friar William De Rubruquis:* chap. 19 (Richard Hakluyt's Edition).

[2] *Fl. c.* 1270–1300.

[3] *The Book of Ser Marco Polo:* Vol. I, p. 226 *et seq.*

[4] *c.* 1330.

[5] *The Journal of Friar Odoric:* chap. 15 (Richard Hakluyt's Edition).

next we hear of him—*c.* 1340—it is as an Abyssinian king.[1]

The real purpose of these missionary expeditions was more ambitious than that of tracing the whereabouts of a semi-mythical king. It was nothing less than the conversion of the Tartar hordes. Although it was not accomplished, the Franciscans achieved a very great deal. Not the least valuable part of their work, from a geographical point of view, was the correspondence arising out of their diplomatic and missionary activities. From the letters of John of Monte-corvino, a Franciscan who left Rome in 1291 and worked among the Tartars for thirty-six years, we derive information, for instance, about such matters as the best overland route to Cathay (he was anxious to have reinforcements for his great task from the Papal See), the kind of vessel employed on the Indian Ocean—' flimsy and uncouth, without nails or iron of any sort, sewn together with twine like clothes ',[2] and the commerce of the Malabar coast. In fact, in his Indian letters[3] we have the first good picture drawn by a Latin Christian, for it was penned before Polo's account. Other medieval ambassadors of the Gospel to leave accounts of the Orient were Andrew of Perugia (who became Bishop of Zaiton), John of Cora, Odoric of Pordenone and—lastly in point of time—John Marignolli.

Unfortunately, the writings both of these men and of their predecessors were not fully appreciated by the educated men

[1] *Vide Mirabilia descripta per Fratrum Jordanum*, edited by Coquebert-Montbret in *Recueil de voyages et de mémoires, publié par la Société de Géographie*, Paris : Vol. IV, 1839, pp. 57–58. By making the Negus of Abyssinia the wealthiest and most powerful monarch in the world, exacting homage from fifty-two kings and at the same time observing certain Christian principles, Jordanus of Severac succeeded in restoring life to the old legend and in directing Western attention to the task of co-operating with the Prester in an attempt to smash the Egypto-Asiatic alliance of Mohammedan power. See also John Marignolli's *Cronica*, *c.* 1355.

[2] *Vide Cathay and the Way Thither :* Vol.III, p. 66.

[3] *Ibid. :* Vol. III, pp. 45 *et seq.*

of their day. Marco Polo was branded as an impostor,[1] and the true stories of Carpini, Rubruck and Odoric were less well known in the later Middle Ages than the fictions and plagiarisms of *Sir John Mandeville*.[2] And yet these works constitute the most important medieval contribution to geography. So expansive is the field of their observations that we cannot attempt here to do more than outline the major achievements. These are remarkable when measured by any criterion, not least by mileage. Down to the year 1245 few, if any, Europeans had travelled farther east than Baghdad.[3] Then in a single expedition an elderly and corpulent friar, by name John of Plano Carpini, penetrated into the very heart of Central Asia, reaching Karakorum, situated 3,000 miles east of the meridian of Baghdad. Ten years later Nicolo and Maffeo Polo, utilizing a similar route, reached the summer capital of the Grand Khan in the hill country behind Peking. Still more remarkable, they returned to tell the tale. But even this feat was overshadowed by that of the younger Polo, Marco, who, in the course of twenty-four years' sojourn in Asia, travelled no less than 30,000 miles, at least 10,000 of which no European had previously followed.

These men, diplomats, traders and missionaries alike, were not only great travellers; they were great observers. Their

[1] *Vide infra*, p. 143. However, Peter of Abano writing *c.* 1300 was courageous enough to call Polo 'the most extensive traveller and most diligent inquirer I have ever known'. *Vide Conciliator differentiarum philosophorum et precipue medicorum*, trans. by H. Yule: *op. cit.*, Vol. I, p. 120.

[2] At least twenty-five editions of *Mandeville's Travels*, as against five of Polo's appeared before 1500. There was no English edition until 1579. With the exception of Roger Bacon's reference (*vide ante*, chap. iv), Rubruck's travels were forgotten until, more than three centuries later, Richard Hakluyt repaired the injustice.

[3] Rabbi Benjamin of Tudela visited the Jewish communities in that city *c.* 1173. *Vide The Itinerary of Benjamin of Tudela*, edited by A. Asher, p. 93 *et seq.* William of Rubruck, however, speaks of 'certain friars Predicants' who visited 'the land of Pascatir'—'a northern region full of woods' (Western Siberia ?) 'before ever the Tartars came abroad'. *Vide Journal, op. cit.*, chap. 23.

narratives abound in descriptions which are self-evidently honest and unadorned, a truly remarkable quality when we recall the indifferent standards of the earlier travel literature, with its disregard for accuracy, and the popularity of the contemporary 'wonder-books'.

From the standpoint of physical geography, these expeditions were particularly notable for their disclosure of the complex mountain system in the heart of Asia. In the writings of the later travellers the Tian Shan, the Pamirs, and the Hindu Kush are recognizably described and given an approximate orientation. Hitherto the classical idea of a single range trending east and west across the continent had prevailed. Even Rubruck did not get beyond it, for when in the neighbourhood of the Tian Shan Mountains he inquired what mountains they were, and from the answer received he understood them to be 'the mountains of Caucasus which are stretched forth and continued on both parts to the sea, from the west unto the east'[1] . . . The feature of Central Asia to attract most attention was the Gobi Desert, which had to be crossed, or at all events skirted, by travellers making their way to Karakorum, the capital of the early Khans. The accounts of it left by Carpini and Marco Polo were scarcely bettered until the nineteenth century. Quoting Carpini, the desert lay

> in that part of the world which is thought to be the most north-easterly . . . In some part thereof it is full of mountains, and in other places plain and smooth ground, but everywhere sandy and barren, neither is the hundredth part thereof fruitful. For it cannot bear fruit unless it be moistened with river waters, which be very rare in that country. Whereupon they have neither villages, nor cities among them, except one which is called Cracurim[2] and is said to be a proper town . . . And albeit the foresaid land is otherwise fruitful, yet it is very commodious for the bringing up of cattle. In certain places it is altogether destitute

[1] *The Texts and Versions of . . . William de Rubruquis*, edited by C. R. Beazley: p. 226.

[2] i.e. Karakorum.

of woods. Therefore the emperor and his noble men and all other warm themselves, and dress their meat, with fires made of the dung of oxen and horses . . . There is never any rain in winter, but only in summer, albeit in so little quantity that sometimes it scarcely sufficeth to allay the dust, or to moisten the roots of the grass. . . .[1]

The great Siberian lowland was obviously 'off the beaten track' of the travellers making for Karakorum and Khan Baliq,[2] so that we shall not expect them to be so exact in their accounts of it. Polo comes as near the truth as any when he says that, beyond the Altai Mountains, there is a plain 'extending to the distance of about forty days' journey . . . [It] borders on the Ocean at its northern extremity.'[3] The inhabitants of this plain 'live upon the flesh of animals . . . they have neither corn nor wine ; and although in summer they derive substance from the chase, yet in winter the cold is so excessive that neither birds nor beasts can remain there'.[4]

On the hydrographical side the contribution of the medieval travellers to the existing body of knowledge was far from negligible, in spite of the absence of useful maps and survey instruments. In the circumstances it is perhaps inevitable that no methodical treatment, say, of the river systems of China was undertaken. But we find it less easy to explain away the absence of any even approximately correct notions of their lay-out and orientation. Even Polo does not appear to have escaped entirely Ibn Batūta's error in supposing that there was

[1] *op. cit.*, p. 108. [2] Peking ?

[3] *The Travels of Marco Polo* (Everyman's Edition) : p. 133.

[4] *Ibid.*, p. 134. Indubitably the finest account extant of the orography of Central Asia comes from the pen of a Moslem whose work—anonymous—has recently become available to western students in a scholarly edition entitled *Hudūd al 'Ālam* (The Regions of the World), translated and explained by V. Minorsky, published by the E. J. W. Gibb Memorial, New Series, Vol. 17, 1937. It is the translator's considered opinion of this work that 'down to the nineteenth century we should look in vain for an equally clear picture of Central Asian ranges'. But the work does not seem to have influenced contemporary or later opinion, Moslem or Christian, to any appreciable extent.

a major north-south system, for in speaking of the port of Zaiton,[1] he says that it is watered by a river which is 'a branch of that which passes the city of Kinsai'.[2] At the same time he realizes that the Yang-tse-Kiang rises in the mountains beyond Sze-chwan,[3] and that the Hwang-Ho has its source in Central Asia.[4] Neither he nor his contemporaries knew anything definite about the southernmost of the three great waterways—the Si-Kiang.

The honour of having made the most important contribution to the hydrography of Asia falls to Rubruck, for his re-discovery of the true character of the Caspian Sea. The River Volga, he tells us,

> runneth into a certain lake or sea, which of late they call the Hircan Sea . . . Howbeit Isidore calleth it the Caspian Sea . . . This sea . . . is compassed in on three sides with the mountains, but on the north side by plain ground. . . . A man may travel round about it in four months. And it is not true what Isidore reporteth, namely, that this sea is a bay or gulf coming forth of the ocean: for it doth in no part thereof join with the ocean, but is environed on all sides with land.[5]

[1] Chuan-chow, near Amoy. 'Were there doubt as to the identity of Zaiton, Abū'l-Fidā's notice would settle it. For he tells us expressly that Zaiton is otherwise called Shanju (Chin-cheu, the name by which Ts'wan-chau [i.e. Chuan-chow] was known to the early Portuguese traders . . .).' —Yule-Cordier: *op. cit.*, Vol. IV, p. 117 note.

[2] Hangchow. *The Travels of Marco Polo* (Everyman's Edition): p. 318.

[3] *Ibid.*: p. 235.

[4] *Ibid.*: p. 272. The Hwang-Ho which, to-day, is barely navigable in its lower reaches owing to the strength of its current, Polo describes as being 'a great river . . . a mile wide and of vast depth' . . . so that great ships can navigate it in comfort. 'At a place in this river, about a mile distant from the sea, there is station for 15,000 vessels.' His description of the Yang-tse-Kiang as 'the largest river in the world . . . upwards of one hundred days' journey [in length] . . . [having] a vast number of other navigable rivers that empty their waters into it . . . and more than 200 [cities and towns] situated upon its banks . . .' (*ibid.*, p. 283) is essentially correct if we discount the superlatives.

[5] *The Journal of Friar William de Rubruquis* (Richard Hakluyt's Edition), chap. 20. Rubruck himself only compassed the northern and western shores of the Caspian. A friend of his, Andrew of Longjumeau, also a

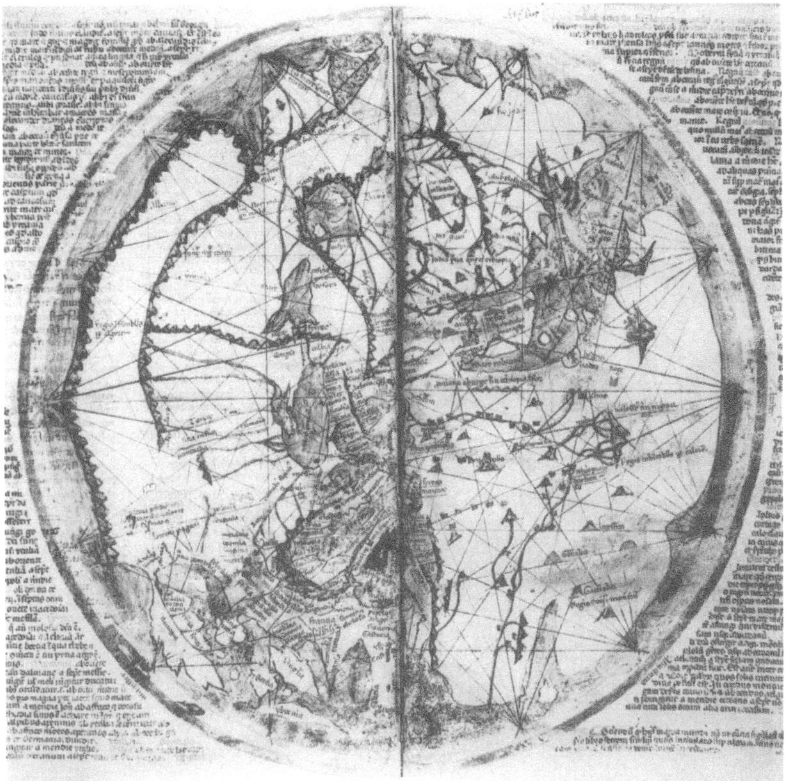

XIII

THE WORLD—ACCORDING TO MARINO SANUTO (1321)

(*Brit. Mus., Add. MSS.* 27,376)

Not only had Isidore held to the contrary opinion, but almost all the writers of geography since the enlightened days of Herodotus. Largely as a result of Rubruck's re-statement of the true case the Caspian begins to be correctly represented on world-maps soon after his time.[1]

Many of the most valuable observations made by these men concern the human, rather than the physical, aspects of geography. Almost without exception the narratives pay considerable attention to Oriental ways of life; the nomadic life of the Tartars, their cart-borne tents, their wandering herds and their food and drink [2]; the idol-worship of the Tibetans and their family cannibalism [3]; the Chinese system of post-stations and the provincial administrative boards [4]; the agricultural economy of the Indians, their religious ritual, ox-worship, caste system and the practice of *suttee* [5]; the pearl fisheries, the cultivation of indigo, ginger and pepper at Quilon, cotton in Gujerat and the huge ocean-going junks which crowded the harbours of the Malabar coast.[6] Aboard one of these many a European traveller sailed from Quilon to Java and Sumatra, even to Indo-China and Cathay, thereby anticipating by more than a century the voyages of the Portuguese caravels.[7] While only Polo has left a detailed account of this maritime commerce with the Far East, several writers speak of the overland traffic and at least one, Francesco Balducci Pegolotti, an agent of the great Florentine house of Bardi, compiled a handbook [8] of the route to be followed in

friar and an envoy from St. Louis to the Great Khan had passed along the southern and eastern shores.

[1] e.g. Marino Sanuto *mappamundi*, *c.* 1320. *Vide* Plate No. XIII.

[2] *Vide The Journal of Friar William Rubruquis*: *op. cit.*, chap. 2 *et seq.*

[3] *Vide* Odoric: *Descriptio Orientalium Partium . . .*, chap. 45.

[4] *Ibid.*: chap. 40.

[5] *Vide Mirabilia Descripta per Fratrum Jordanum . . .*: *op. cit.*, pp. 42 *et seq.*

[6] *Vide* Odoric: *op. cit.*, chap. 16, and Marignolli, in Yule-Cordier: *op. cit.*, Vol. II, pp. 309–94.

[7] *Vide* Odoric: *op. cit.*, chap. 20. Polo speaks of 'merchants from Manzi (i.e. South China) trading at Quilon'. *Travels*: *op. cit.*, p. 377.

[8] *Libro di divisamenti di paesi.*

going from the Levant to Peking. His observations include the stages of the journey, the mode of transport appropriate to each stage, the time normally taken, the most suitable articles of commerce with the Tartars and Chinese and, not least, the profits likely to accrue from the enterprise. He assures merchants intending to make the journey that 'the road you travel from Tana [1] to Cathay is perfectly safe, whether by day or by night, according to what merchants say who have used it . . .' [2]

For the greatest wealth of information regarding the economic activities of Eastern peoples and their cultural landscape, we are indebted to Polo. His *Travels* are, in fact, so rich in allusions to these topics that judicious selection from them is embarrassing. To the student of geography not the least interesting passages are those where he alludes to the extraordinary density of Chinese population and the measures employed to cope with it. So great is it that 'no spot of earth is suffered to lie idle that can possibly be cultivated '.[3] Fortunately, the productivity of the soil is such that the three main grain crops: 'rice, panicum and millet . . . yield, in their soil, an hundred measures for one'.[4] Doubly fortunate are they in possessing 'a sort of black stone which they dig out of the mountain where it runs in veins', for when it is lighted 'it burns like charcoal and retains the fire much better than wood . . . It is true that there is no scarcity of wood in the country, but the multitude of inhabitants is so immense . . . that the quantity could not supply the demand. . . .' [5] Much as the rural density of the population impressed Polo, he was even more astounded at the astronomical proportions of the cities. Occasionally his astonishment takes a rather extravagant turn, as in the case of Kinsai—'the celestial city, [a name] which it merits from its pre-eminence to all others in the world, in point of grandeur and beauty, as well as from its abundant

[1] Situated near the mouth of the R. Don.

[2] *Vide* Yule-Cordier: *op. cit.*, Vol. II, pp. 279–308.

[3] *Travels:* (Everyman's Edition), p. 209.

[4] *Ibid.:* p. 209.

[5] *Ibid.:* p. 215.

delights which might lead an inhabitant to imagine himself in paradise '.[1] But there was some excuse for it, if, as he claims, the city was one hundred miles in circuit,[2] contained ten principal market places each of which was attended three days a week by ' an assemblage of from forty to fifty thousand persons '[3] and housed 1,600,000 families.

In the light of such facts, it is not surprising to read that, then as now, famine was a periodic menace to the stability of life in China. To grapple with it the Khan sent out his commissioners everywhere to

> ascertain whether any of his subjects have suffered in their crops of corn from unfavourable weather, from storms of wind, or by locusts, worms, or by any other plague ; and in such cases he not only refrains from exacting the usual tribute of that year, but furnishes them from his granaries with so much corn as is necessary for their subsistence, as well as for sowing their land. With this view, in times of great plenty, he causes large purchases to be made of such kinds of grain as are most serviceable to them, which is stored in granaries provided for the purpose in the several provinces, and managed with such care as to ensure its keeping for three or four years without damage.[4]

Another thoughtful undertaking of the reigning Khan's, and one which has had a permanent effect upon the Chinese landscape, was the planting of large, tall trees along the high roads. Besides giving much-valued shade in summer-time, they served ' to point out the road when the ground is covered with snow '.[5] In the course of his long journeyings up and down the country no doubt Polo had continual reason to be grateful to the Khan for this provision, as also for the post-houses situated ' upon every great high road, at the distance of twenty-five or thirty miles '.[6] Not a little to his astonishment, Polo found that ' even in mountainous districts, remote

[1] *Travels : op. cit.*, p. 290.

[2] Even supposing these miles to be Chinese—1 *li* equals 1,900 feet approx.—the circumference is still enormous measured by medieval European standards.

[3] *Travels : op. cit.*, p. 293.

[4] *Ibid.* : pp. 212–13.

[5] *Ibid.* : p. 214.

[6] *Ibid.* : p. 206.

from the great roads, where there were no villages, and the towns are far distant from each other, his majesty has equally caused buildings of the same kind to be erected, furnished with everything necessary . . .'[1]

Coming, as he did, of a merchant family, Polo may perhaps be forgiven for the strong emphasis he is inclined to give to Oriental manufactures and articles of trade. All told, he mentions over eighty different marketable commodities. Prominent among them is pepper, one of the most coveted spices in the medieval Western world, the possession of a few pounds of which spelt riches. We can well imagine Polo's thoughts, therefore, when he learned that the *daily* consignment of pepper from Zaiton to the town of Kinsai alone was 'forty-three loads, each load being two hundred and forty-three pounds'.[2] Elsewhere, in connexion with the trade of Zaiton, he surmises that 'the quantity of pepper imported there is so considerable that what is carried to Alexandria to supply the demand of the western parts of the world is trifling in comparison, perhaps not more than the hundredth part'.[3] Among the many other articles referred to by Polo, porcelain ware receives more than its share of attention, though this is, of course, hardly surprising in a Venetian. The method of its manufacture is described by him in detail, as also is the manufacture of wine, paper, indigo, woollen cloth and carpets,[4] but in a work of this compass it is not possible to do more than make a passing reference to them. Fortunately the *Travels* are available in so many editions that the study of them at first hand presents no difficulty.

But enough has been said to indicate the character and the quality of these Asiatic narratives. Their authors were essentially reporters, content for the most part to describe what they had seen or been told. In some instances it would seem

[1] *Travels: op. cit.*, p. 208. [2] *Ibid.*: p. 297. [3] *Ibid.*: p. 317.

[4] Tea is a notable absentee, notwithstanding the fact that Polo crossed the tea-growing terraces of Fokien. As a matter of fact none of the medieval European travellers in China mention tea.

that they did not describe *all* that they had seen. Thus an analysis of Polo's *Travels* suggests that he withheld many of the more spectacular facts of his Oriental experiences, for fear, perhaps, of courting the scorn of his sophisticated compatriots.[1] There is, for instance, no mention of the Chinese wall, although it is quite clear that Polo travelled within sight of it, if not actually on it for a considerable distance.[2] Similarly nothing is said about the compressed feet of the Chinese women, or the use of fishing cormorants, printing presses and the peculiar characteristics of Chinese writing. Only Odoric, among the Christian writers of the Middle Ages, refers to the compressed feet of the women, or to the cormorants.

Seldom are these 'reporters' disposed to examine the implications of their facts, or question the *bonâ fides* of their informants. Of scientific ideas, there is practically no trace. Polo solemnly assures us that a certain island in the Northern Ocean lies 'so far to the north that the polar constellation appears to be behind you and to have in part a southerly bearing',[3] and in another place he estimates the latitude of a locality by counting the number of cubits high the Pole Star is above the horizon.[4] As for the relationship of the new knowledge to the old beliefs, it is mostly avoided. However, Rubruck, in speaking of the Cathayans, surmises that they are the Seres of the Ancients,[5] and John Marignolli, in describing the confines of Europe, postulates a 'White Sea' beyond

[1] Jacopo of Acqui, a contemporary of Polo's, writing in a somewhat obscure chronicle called *Imago Mundi* declares that 'what he (i.e. Polo) told in the Book was not as much as he had really seen, because of the tongues of detractors who, being ready to impose their own lies on others, are over-hasty to set down as lies what they in their perversity disbelieve, or do not understand. And because there are many great and strange things in that Book which are reckoned past all credence, he was asked by his friends on his death-bed to correct the Book by removing everything that went beyond the facts. To which his reply was that he had not told one-half of what he had really seen.' *Vide* H. Yule-H. Cordier: *The Book of Ser Marco Polo . . .*, Vol. I, p. 54.

[2] *Vide* H. Yule-H. Cordier: *op. cit.*, Vol. I, pp. 292–4.

[3] *Travels: op. cit.*, p. 134. [4] *Ibid.*: p. 379.

[5] *The Journal of Friar William De Rubruquis*, chap. 28.

Hungary and hints that there is no need to verify the fact for he had learned by 'sure experience that fully half the surface of the Ocean was unnavigable', adding that 'God did not wish the human race to sail round the world'.[1]

For all the awareness he appears to manifest of the controversies of ancient and medieval geographers, Polo may well have been entirely ignorant of them. Thus, the fact that he found several well-populated islands in the 'uninhabitable' Torrid Zone, e.g. Madagascar, Zanzibar, Java and Sumatra, occasions him no surprise and stimulates him to no further inquiry. His attitude towards the unknown and the problematical is admirably summed up for us in the passage where he speaks of the 'eastern ocean' with its 7,448 islands.[2] 'We shall cease,' he says, 'to treat further of these countries and islands, as well on account of their lying so far out of the way, as of my not having visited them personally . . .'[3] Even Japan (Zipangu)—'an island in the eastern ocean, situated at the distance of about 1500 miles from the mainland'—is dismissed in two short chapters, and these are chiefly concerned with the description of a Tartar expedition against the country and not with the country itself.[4]

The immediate results of these explorations upon geographical theory and cartography were as paltry as those for commerce were profound. Among the factors contributing to this misfortune, the unreal character attributed to the travel literature was most probably an important one. Moreover, we must not forget the imperfect nature of publication before the invention of the printing-press, the reluctance of the age to abandon traditional beliefs, and the absence of rigid scientific principles in the geography of the time. Whatever the true explanation, contemporary chroniclers are extremely reticent about the heroes of their age. Sir Henry Yule only succeeded in finding five references to Polo in the whole of the fourteenth century, and there are scarcely more than this

[1] *Vide* H. Yule-H. Cordier: *Cathay and the Way Thither*, Vol. III, p. 246.

[2] The precise number varies from manuscript to manuscript.

[3] *Travels: op. cit.*, p. 330. [4] *Ibid.*: pp. 323 *et seq.*

number in the writings of the early fifteenth century.[1] Although there was a latin copy of his *Travels* in a Boulogne library from 1320 onwards, Pierre d'Ailly, the famous Cardinal and scholar, who lived in the neighbouring town of Cambrai, *c.* 1410, never once cited the name of Polo, and did not even appear to know the name of Cathay. Again, although Prince Pedro, Prince Henry the Navigator's elder brother, returned from Venice in 1428 with a copy of the *Travels* and a *mappamundi*,[2] neither Zurara nor Galvão (who does tell how the Navigator was 'much furthered' in his discoveries by the map) makes any reference to the utility of the book. In fact, almost the only text-book of geography, prior to the great age of discovery, to allude to the Indian Ocean in a way suggesting that the author was not altogether ignorant of recent discoveries is the *Tractatus optimus super totam astrologiam* of Bernard of Verdun.[3] In his discussion of the numerous islands located in that sea, he speaks of one—the greatest—lying remote from the mainland of Asia and having a circumference of some 3,000 miles, a figure which agrees exactly with that given by Polo for Madagascar.

The world-maps of the period are little better than the text-books, for only occasionally do they show even a few of the new place-names and physical features. All the Hereford Map, *c.* 1280, shows that is post-Dark Age in ancestry is the solitary name of Samarkand, and the Marino Sanuto *mappamundi*, *c.* 1320, shows only the Caspian Sea corrected to agree with the findings of Rubruck and Longjumeau. The Catalan Atlas of 1375 goes somewhat further in its adoption of the new knowledge, but even here it is not always easy to understand the view of the travellers' itineraries taken by the compiler. His names, moreover, are often unrecognizable, likewise his outlines: e.g. Malay lacks its peninsula. At the

[1] It was not until the Columbian era that the work came into its own. Columbus himself would seem to have consulted it freely. *Vide* autograph notes on the printed edition of the *Travels* in the Columbina Library, Seville.

[2] *Vide ante*, chap. v.

[3] *Fl. c.* 1300.

same time the chief regions of India and China and several of the more important East Indian Islands, notably Sumatra, are delineated in fairly close correspondence with the indications of the Polan narrative.[1] As for the distinguished fifteenth-century cartographers Fra Mauro and Martin Behaim [2] neither of them took the trouble to lay down the itineraries (in as far as they allow of reconstruction), being content to combine the erroneous designs of incompetent older authorites. Paradoxically, the only map which does give all or almost all the information found in the various narratives dates from the sixteenth century, by which time it had become obsolete.[3]

In face of such apathy, it is small wonder that the glorious achievements of the Great Age of Medieval Exploration were short-lived. With the death of John Marignolli, Europe's last ambassador to Peking, the Age drew to its close. Twenty years later, the Black Death began to scourge the West, reducing the population of Europe almost by half, disorganizing society and temporarily paralysing the energies of her peoples. Twenty years later still, the Mongol dynasty which for some time past had shown signs of weakening, fell before the rising power of the Mings. The new rulers of China reverted to the traditional policy of the country and henceforth kept foreigners at a respectful distance. Consequently, missionaries and merchants vanish from the Oriental scene.[4] Darkness once more covers the face of the East, and both Manzi and

[1] H. Cordier is of opinion that 'the cartographer's knowledge of Eastern Asia is drawn entirely from Marco Polo'. *Vide* H. Yule-H. Cordier: *Cathay and the Way Thither*, Vol. I, p. 302.

[2] *Vide infra*, chap. viii.

[3] *Vide* A. E. Nordenskiold: The Influence of the Travels of Marco Polo on Jacobo Gastaldi's Maps of Asia, *Geographical Journal*, 1899, pp. 296–406.

[4] The only European travellers of this new era whose works survive, Clavijo and Schiltberger, did not penetrate farther east than Samarkand (*Vide* Hakluyt Society editions of their works). But it is evident from their writings that a great commerce continued between China and Russia, for while Clavijo was at Samarkand a caravan of 800 camels arrived there from China with silk, gems, musk and rhubarb which were bartered with Russian merchants for linens and furs.

Cathay are lost to European view. Even the memory of them becomes dim.

But we should be wrong if we supposed that these calamities nullified the achievements of the Asiatic explorers. The immediate results upon geographical theory may have been small. The missionary results were certainly not startling, but the wealth of rumour and story was not quickly dissipated, nor was the memory of the lucrative trade in spices and precious metals soon forgotten. On the contrary, these things served to stimulate some, at any rate, of the disinherited peoples to find alternative routes to the 'Indies'.[1] So great, indeed, was the psychological effect of the discovery of the Far East that it is scarcely an exaggeration to say that it did more than any other medieval happening to change the *Weltanschauung* of European civilization. While it was not until the fifteenth century that the 'Terrestrial' era of human history was superseded by the 'Oceanic', a reading of fourteenth- and fifteenth-century literature makes it obvious that the transition from the outlook which accounted all the world as tributary to the Mediterranean to that which admitted no limits to the *oikoumenē*, had its origins at least two centuries before the time of Columbus.[2] Fifty years of Cathay had wrought a bigger change in the outlook of the Western world than a whole cycle of medieval Europe.

[1] The name 'Cathay' is only rarely found in the text-books and maps of the fifteenth century.

[2] Few books express better the dissatisfaction and restlessness of a good European of the early fourteenth century at being cooped up nautically in the Mediterranean than Marino Sanuto's *Liber secretorum fidelium crucis super terrae sanctae recuperatione et conservatione* (written *c.* 1306–20).

CHAPTER VII

PHYSICAL GEOGRAPHY IN THE MIDDLE AGES

NO community, no matter how slight its hold upon the material foundations of life, or how remote it may be from the centres of civilization, can remain entirely uninterested in the surrounding physical world. If its people are valley-dwellers, following a settled habit of life, they are of necessity interested in the behaviour of the river which fertilizes their valley. After all, asked the Egyptians, why should the Nile overflow its banks every summer just when the land is languishing under a burning sun? If its people live by the sea, they can scarcely refrain from speculating as to why its waters are always in motion, or what is the cause of their ebb and flow, and why they are salt. As for people unfortunate enough to live within range of seismic disturbance, they need no encouragement to inquire into the cause of such sinister visitations. In short, we may say that interest in matters of this kind is elemental and almost as universal as man. Further, it has a habit of persisting even during those dark eras in the history of the race when worldly interest is small. This was certainly the case in the Middle Ages, for while it is the exception to find an intelligent curiosity concerning purely cosmographical topics, many a writer makes thoughtful observations on the behaviour and causation of natural phenomena. The primary reason for this circumstance rests, no doubt, with the impression made upon all peoples by such things as tides, thunder, lightning and earthquakes. Spectacular and unpredictable, for the most part—at all events by unscientific men—they command the attention more readily even than 'travellers' tales'. A contributory factor may be cited in this particular case, namely, the greater latitude allowed by the early Church in purely physical matters. Whereas orthodoxy prescribed the acceptance of very definite ideas concerning the shape, movement and peopling of the earth, it had much less to say upon meteorology, climatology and oceanography—doubtless because the Bible was less informative.

This interest in the physical universe was, moreover, not confined to the intellectual giants of the thirteenth century; it is even apparent in the writings of some of the early Fathers. Ambrose, among others, argues the connexion between tides and the lunar cycle. Augustine theorizes about the origin of springs and rivers. Bede and Dicuil debate whether volcanoes are burning mountains or vents for deep-seated terrestrial fires. Isidore explains why the seas do not overflow their banks, by pointing out that water is constantly being removed into the air and on to the land. It is true that these observations mark no development upon earlier ideas, that the Ancients had said it all before, and that, in most cases, they had said it better, but at least they betoken an interest in the pursuit of learning for its own sake—an uncommon quality in the scholarship of that age. At the same time it has to be admitted that, even in secular matters of this kind, the scholarship of the Dark Ages was not completely free from the shackles of Biblical exegesis.[1] When the Scriptures did speak of physical things, 'let God be true and every man a liar'.

The Biblical topic to cause as much confusion as any in the minds of the Church Fathers was the statement in the Creation story that 'God made the firmament and divided the waters which were under the firmament from those which were above the firmament. . . .'[2] Now there was nothing novel

[1] A common practice was for writers to treat such subjects in the order in which they are first mentioned in Genesis. Isidore in his lesser known treatise on physical phenomena, entitled *De Natura Rerum* deals, first of all, with time periods (i.e. chronology), then the heavens (i.e. astronomy, to which are appended a few chapters on meteorological matters), followed by the sea (oceanography) and the dry land (physiography).

[2] *Genesis* i. 7.

about belief in waters above the firmament; it is found in the cosmogonies of the ancient Egyptians and the Persians as well as in Chaldean literature, where it is also associated with belief in a dome-shaped heaven. But it was very far removed from the views of the universe propounded by Aristotle and his followers. In order to 'rationalize' the belief, some of the Fathers went to the extent of imagining mountains at the extremity of the firmament and holding that the waters were contained in the hollows and valleys between them.[1] Others thought that the waters were not to be taken as literal, but merely as being water-vapour, while some, including Jerome, Ambrose and Bede, held that the waters were crystal. Augustine was wiser than many in his generation and simply contented himself with collating the various views held on the subject.

Where scientific arguments and evidence were lacking, the Fathers commonly fell back upon broad religious principles, maintaining that God's relation to Nature is immediate and not conditioned by any observed rules. And so we find men denouncing 'the vain and curious desire of investigation palliated under the name of knowledge and science', on the ground that it is apt to lead one into 'searching through magic arts to the confines of occult science'.[2]

But even among the orthodox there was considerable diversity of opinion on physical matters. Thus the statement in the Psalter to the effect that God had established the earth upon the seas[3] was variously construed to mean that the earth was actually floating on water (this was the orthodox view), that the land was established *beside* the seas (the word 'super' in the Vulgate being given the force of 'over against'), that the land rises to a higher level than the sea, and that the statement was allegorical and never intended to be taken literally. As late as the fourteenth century the problem was still exercising men's minds. The anonymous writer of an

[1] *Vide* Gregory of Nazianzus: *In Hexaemeron Liber* in Migne: *Patrologia Graeca*, Vol. XLIV, cols. 89–90.

[2] Augustine: *Confessions:* x. 35. [3] *Psalm* xxiv. 2; cf. *Psalm* civ. 3.

encyclopaedia entitled *Lumen Animae* or *Book of the Moralities of Great Natural Things* solves it to his own satisfaction by granting that in actual location water is above the earth but that virtually it supports the earth, because without water the earth would dry up and be dissolved into powder. Any suggestion that the Biblical text might not be verbally accurate appears to have been entirely alien to the mentality of the age. All the same, a mere passing acquaintance with the literature of the age makes it quite clear that the holding of orthodox Christian faith and the enlargement of physical knowledge were by no means incompatible. During the medieval period real progress was registered in almost every department of the physical sciences and Roger Bacon, the greatest of all medieval scientists, would have been the first to deny the imputation of unorthodoxy. This progress is a tribute in no small degree to the wisdom of scholars in refusing to fight over the bones of theological contention which they could not digest.

To mark each step of this progress would be a long and tedious task, even supposing that it were practicable—which is highly debatable—since the spread of new ideas in the Middle Ages was slow and at best sporadic. The most we can hope to do within the compass of a single chapter is to indicate the main contributions to physical studies made through the centuries.

METEOROLOGY AND CLIMATE

At the commencement of our period the ideas in circulation about the atmosphere were classical in origin and were derived, almost exclusively, from Pliny's *Natural History* and Seneca's *Natural Questions*. Of the two works, Pliny's enjoyed the greater popularity, as may readily be seen by reading Isidore or Raban Maur on such subjects as winds and climatic zones. Departures from the normal, when they occur, are more apparent than real, for they are usually made in favour of some later pagan plagiarist of Pliny, e.g. Solinus or Mela, for the very thought of originality is foolishness alike to pagan and priest at this era. The only kind of novelty to be suffered is an eye-witness observation, inserted parenthetically, or a denunciation of current magical practices. In the former category, we have Bede's guide to weather forecasts—based, no doubt, upon observations made in the then rural Jarrow.

> If the sun is blurred upon rising, or hidden behind clouds, then [he contends] a rainy day is indicated. If it is red, it will be fair.[1] If it is pale, it will be stormy; if it seems concave, gleaming so that in the centre it sends rays to the north and the south, then the weather will be wet and windy . . . Lightning from the north or thunder from the east portends a storm, while wind from the south portends heat . . . Again when at night water shines upon the oars, there will be a storm. And when dolphins leap frequently from the waters in which they are carried, then a wind will arise and broken clouds open up the sky.[2]

But lest we should imagine that observations of this sort are typical of the 'popular' meteorology of the time, we need only refer to a treatise by Agobard, a ninth-century Archbishop of Lyons, entitled *Liber contra insulsam vulgi opinionem de grandine et tonitruis* (Against the Absurd Opinion of the Vulgar touching Hail and Thunder.)[3] In this the writer inveighs against charlatans who claim that they could control the weather, produce storms and hail at will. 'The wretched world,' he declares, 'lies now under the tyranny of foolishness; things are believed by Christians of such absurdity as no one ever could aforetime induce the heathen to believe.' The general medieval view, sanctioned by Popes and shared by poets, painters and priests, was scarcely less foolish however. 'It is a dogma of faith,' says St. Thomas Aquinas in his all-authoritative *Summa Theologica*, 'that demons can produce winds, storms and rain of fire from heaven.'[4] By the side

[1] cf. *Matthew* xvi. 3. [2] *De Natura Rerum*: cap. 36.

[3] *Vide* J. P. Migne: *Patrologia Latina* . . ., Vol. CIV (also R. L. Poole: *Illustrations of the History of Mediaeval Thought*, pp. 40 *et seq.*).

[4] St. Paul speaks of the Devil being 'the prince of the power of the air'. *Ephesians* ii. 2.

of such extravagances, Isidore's physical explanations are the work of a modern rationalist! Rain arises from an exhalation from land and sea, which being carried aloft falls in drops on the lands, being acted upon by the sun's heat, or condensed by strong winds.[1] Lightning is caused by the collision of clouds,[2] thunder by their bursting;[3] the rainbow, by the sun shining into a hollow cloud.[4]

As the centuries unfold, the bias towards personal observation and experiment becomes increasingly more apparent. The works of Pliny and Seneca are still being read and referred to, but their authority no longer passes unchallenged. By the twelfth and thirteenth centuries almost every physical work exemplifies, in one way or another, this renewal of intellectual vitality. In the forefront stand the *De Philosophia Mundi* and *Dragmaticon Philosophiae* of William of Conches. An early predilection for physics led him to examine the views of Seneca whom he thereafter held in high regard. However, his was no mood of servile acquiescence. Among other things, William had very definite ideas about the circulation and distribution of thermal conditions in the atmosphere. Recognizing that its density and humidity decrease progressively with altitude, he deduced that the sun's influence on the denser air of low altitudes is far more potent than it is on the rarer strata above.

Although heat is emanated from the sun, which in the nature of things is warm, it can warm nothing until it becomes mingled with humidity, which it transforms into its own nature. Hence in the mountains, because the air is rarefied owing to its very great distance from the earth, it cannot be warmed by the sun; instead, as a result of its lightness, it is driven here and there and consequently cooled. In the valleys, on the other hand, because the air is dense and, therefore, almost stationary, it is quickly warmed and becomes hot.[5]

[1] *Etymologiae*: XIII, 10, ii.
[2] *Ibid.*: XIII, 9, i.
[3] *Ibid.*: XIII, 8.
[4] *Ibid.*: XIII, 10, i.
[5] *De Philosophia Mundi*: IV, 5 (in J. P. Migne: *Patrologia Latina* . . ., Vol. CLXXII).

But if the atmosphere became increasingly rarefied with rising altitude, how was it that the upper levels are so often more cloudy than the lower levels? This question William of Conches rightly answers by arguing that clouds are not composed of air of greater density than the surrounding parts of the atmosphere, but of 'water vapour which rises up from the valleys (owing to its greater warmth) and is condensed into clouds and snow upon coming into contact with the cooler air of the upper atmosphere'.[1] Furthermore, he appreciated the fact that this process was frequently productive of rain.[2] He also observed (what is by no means as impossible as it seems[3]) that little frogs and fishes had been known to fall with raindrops during whirlwinds and deduced from this fact that the water-vapour must have been whipped off the surface of lakes, swamps or rivers.[4]

To the medieval scholar the subject of rain was never very far removed from that of the Deluge (of the Bible). Flooding on a normal scale he could readily understand, although it need not be supposed that he understood the phenomenon of long spells of wet weather to which it was due; but how could the whole of the *oikoumenē* be inundated at one and the same time? and obviously the waters had flooded

[1] *De Philosophia Mundi*: IV, 5. Cf. *Image du Monde*: chap. 15. 'Of therthe groweth the rayn and the clowdes also, as of a cloth that is weet and shold be dreyd by the fyre; thenne yssueth like a smoke or fumee and goth upward. Who thenne helde his hande over this fumee, he sholde fele a vapour which sholde make his hande moyst and weet; yf it dured longe he sholde appertly knowe that his hande were alle weet, and that water sholde droppe and falle therof. And thus I saye to yow that in this maner growe ofte the clowdes and raynes.' (Caxton's translation.)

[2] *De Philosophia Mundi*: III, 4.

[3] *Vide* E. W. Gudger: Rains of Fishes, *Natural History* (Journal of the American Museum of Natural History), Vol. XXI, 1921, pp. 607–19.

[4] *De Philosophia Mundi*: III, 6. 'Cum enim . . . aqua vento sustollitur, contingit quod ranunculos et pisces secum elevat, quibus ex naturali grandine descendentibus, stupent ignorantes.'

the whole earth, for, said Isidore (and his numerous followers agreed)

we observe the proof of this even at the present time in the stones which we are wont to go and see in the distant mountains, which have mingled in them the shells of mussels and oysters and besides are often hollowed by the waters.[1]

Several hypotheses were put forward. Peter Abelard cited the theory—which he did not hold—that it was the purpose of the mysterious 'waters above the firmament' to provide these waters.[2] Peter Comestor[3] and Gervase of Tilbury[4] were of the opinion that they came partly from the earth beneath and partly from the air above, a view that is in conformity with the statement in the seventh chapter of *Genesis* to the effect that 'all the fountains of the great deep [were] broken up and the windows of heaven were opened'. Another theory based upon the express words of Scripture and also mentioned by Gervase is that no rain had fallen upon the earth until the Deluge came.[5] The accumulated water-vapour in the atmosphere might be held, not unreasonably therefore, to have supplied the necessary volume of rain.

In dealing with the question of winds and atmospheric circulation in general, the writers of the later centuries were not so hampered by the *dicta* of theology. Some of their theories, in fact, are quite remarkable for their novelty. Thus, William of Conches submits[6] that the cause of the planetary wind system is bound up with ocean currents. He believed that in equatorial waters there are two counter-flowing currents each of which bifurcates at the extremity of the habitable earth. The resulting collision caused the generation

[1] *Etymologiae*: XIII, 22, ii.

[2] *Expositio in Hexaemeron, De Secunda Die*, in Migne: *Patrologia Latina*, Vol. CLXXVIII, col. 743–4.

[3] *Historia Scholastica, Liber Genesis*: cap. 34, in Migne: *op. cit.*, Vol. CXCVIII, cols. 1045–1722.

[4] *Otia Imperialia*: I, 24.

[5] *Genesis* ii. 5–6.

[6] *De Philosophia Mundi*: III, 15.

of Zephyr and Eurus, two of the cardinal winds.[1] At the same time, the collision causes one branch of the current to flow northwards and the other to flow southwards. These branches, he argued, would collide at the Poles, their impact disturbing the atmosphere and giving rise to Boreas and Auster, the North and South winds. If by any chance one of the currents flowed more strongly than its opponent, then the point of collision would be pushed away from the pole. This displacement (of the point of collision) would explain, according to William, the blowing of collateral, or oblique, winds, e.g. north-easterly and north-westerly winds.[2] Impossible as it may be in itself, this theory is not without interest because 'it shows that William understood that a broad system of atmospheric circulation is possible, and assigned to it . . . a purely physical cause'.[3] The fact that he supposed the winds to be an effect rather than a cause of ocean currents is not nearly as important as the fact that he attempted to construct a reasoned system of causation.

As to the original cause of the planetary movement of the air, nobody was able to offer a really satisfactory explanation. Although many a person echoed the query, put by the *Nephew*

[1] Cf. *Image du Monde*, II, 29, which is taken from *De Philosophia Mundi*: 'The wyndes renne round aboute therthe oftymes and entre counte and mete in som place so asprely that they ryse upon heyght in such wyse that they lyft up thayer on hye. And thayer that is so lyft and taken from his place remeveth other ayer in suche facion that it reterneth as it were afterward, and gooth cryeng and brazeng (= braying) as water rennyng; for wynde is none other thyng but ayer that is meuyd so longe tyl his force be beten down with the stroke. Thus come ofte clowdes, raynes, thondres and lyghtnyngs. . . .' (Caxton's translation.)

[2] His argument, for what it is worth, runs as follows: 'Si in duabus refluxionibus ad septentrionem vergentibus, Boreamque praedicto modo generantibus contingat, quod orientalis ex aliquo accidente sit velocior, ultra medium septentrionem et occidentem occurrit, nasciturque ventus collateralis, inter Boream et Zephyrum. Si vero occidentalis, ultra medium septentrionalis orientali occurrit, fit ventus collateralis inter Boream et Eurum.'—*De Philosophia Mundi*: III, 15.

[3] Wright: *Geographical Lore of the Time of the Crusades*, p. 173.

in Adelard's *Quaestiones Naturales* (*LX*), the *Master's* solution was seldom bettered.

Just as that which through its abounding in fire is most light —I mean the ether—holds a position near the fire, so the wind, being weighed down by the heaviness of moisture, has its abiding place near the water ; the former is raised by its rarity, the latter depressed by its weight : one is elevated by heat, the other weighed down by the coldness inherent in moisture. Consequently, as it is bound to move and cannot rise, it courses round the earth and troubles our peace.[1]

The extraordinary force of wind and its destructive power were commonly explained on the assumption that

no small amount of air is imprisoned in the bowels of the earth. . . . As time goes on, however, either its own strength giving it the means or the earth's established course opening a road to it, it manages to burst forth, and the longer it has remained shut up, the more violent will be its exit. If in addition, it becomes of a stout constitution in the matter of density thro' an uncertain watery humour, this stoutness combined with swiftness produces propulsive force.[2]

On the climatic import of these physical movements our writers speak with more orthodoxy than originality. For, both in Antiquity and in the Middle Ages, the portions of the earth's surface near the Equator were universally regarded as being hot and dry, the portions near the poles as cold and damp. Upon this schematic arrangement few people were able to improve, beyond the suggestion volunteered by William of Conches, that the temperature varies in direct relation with distance from the Equator—'Aequaliter vero distans, aequaliter est temperata '.[3] But there was not the same general agreement concerning the climates of lands in the same latitudes. Bartholomew Anglicus, for instance, supposed that the western regions of the habitable earth were cold and damp, the eastern warm and dry. William of Conches, on

[1] *Quaestiones Naturales* (H. Gollancz's edition) : p. 148. [2] *Loc. cit.*
[3] *De Philosophia Mundi :* IV, 3.

the other hand, transposed the characteristics to cold and dry and warm and damp. From this over-simplified conception of the distribution of climatic qualities, William then deduced that

> places cut off from the north winds by mountains have dry, warm conditions and are good for winter residence, but very undesirable in summer. The opposite is true of places on the north sides.[1] Similarly, places facing the east and shut off from other winds are warm and damp in spring, which is a pleasant season, but dry and cold in autumn.[2]

The converse, he avers, is true of places with a western exposure.[3]

There may be nothing very startling about these observations, but at least they are an advance upon the earlier pseudo-classical writings. In some ways they are more modern in temper than some that came immediately after; William and his contemporaries were writing before the full recovery of Aristotle's scientific works. The translation of these into Latin was not entirely an unmixed blessing for physical studies, for though they were greatly superior to those hitherto held in authority, they had the effect of restraining independent scientific inquiry. The more cautious, subservient spirit induced by Aristotelianism is amply demonstrated by such a work as the *De Proprietatibus Rerum* of Bartholomew Anglicus, written approximately one hundred years after Conches' *De*

[1] *De Philosophia Mundi*: IV, 6. 'It was for this reason that the Ancients used to make their dining-rooms southerly and northerly, so that they might dine in the southerly room in winter, and in the northerly one in summer.'—*Ibid.*

[2] *Ibid.*

[3] The concepts of the four elements (earth, water, air and fire) and their properties (dry, damp, cold, hot) were fertile in every branch of medieval natural science. It is by reference to them that the seasons (and often the winds) were commonly explained. 'The spring is composed of moisture and heat; the summer, of fire and dryness; the autumn, of dryness and cold; the winter, of cold and moisture.'—*Vide* Isidore: *De Natura Rerum*, VII, 4.

Philosophia Mundi.[1] It is true that the *De Caelo et Mundo* and other standard works of Aristotle come in for criticism, but it is always highly respectful. The note of originality sounded more than once in Conches' writing is all but missing here. Bartholomew's discussion of lightning is typical.

> There are different views concerning the nature and origin of lightning. Empedocles says that lightning is fire hidden in clouds and is generated by the sunbeams. But Aristotle says that this view is wrong, for, if it were true, lightning should come out of all clouds because some sunbeams are hidden in every cloud. And Anaxagoras says . . . Another says . . . Yet others say . . . Aristotle says . . . and Aristotle means. . . . Gregory says . . . as Bede declares . . .[2]

And so the tale continues. Aristotle is quoted no less than thirty-four times in the section on the atmosphere, Bede twenty-four times, Isidore twenty, Constantine thirteen, Gregory five, others thirteen. Bartholomew's own opinion —in as far as anything is his own—favours Aristotle's. He puts lightning and thunder down to the 'moving and shaking hither and thither of hot and dry vapours fleeing from their contrary qualities. Being beset and constrained and smitten on every side, they are thereby set on fire and on flame. . . .'[3] But the restraint was only temporary and never more than partial. It had little or no power over men like Robert Grosseteste,[4] Roger Bacon or Albert the Great.[5] With the result that their physical observations commonly transcend both those of William of Conches and Bartholomew Anglicus. At the same time their superiority is one rather of argument than of achievement, asserting itself in its relative freedom from inconsistency of thought and in the logic of its deductions rather than in the establishment of new hypotheses. It is largely on such grounds as these that Albert's climatic observations, for instance, have won universal admiration and are

[1] i.e. mid-thirteenth century.

[2] *De Proprietatibus Rerum :* XI, 14.

[3] *Ibid. :* XI, 13.

[4] Bishop of Lincoln in the mid-thirteenth century (d. 1253).

[5] *Vide ante*, chap. iv.

entitled to the distinction of being the first attempt at a comparative geography. His original contributions are few and consist chiefly of an elaboration of the part played by the sea, mountains, vegetation and the soil on terrestrial life, more particularly upon human beings. The classical masters of science had made all climatic and bio-geographical variations depend rigidly upon latitude. To Albert and his contemporaries belongs the credit of having established the influence of elevation and aspect, even though they grossly underestimated the complexity of such influences, as, for example, where Albert correlates southern, northern, western and eastern aspects with comparative warmth, cold, wet and dryness respectively, and where he tries to trace the influences of latitude upon wheat.[1]

A sign of the times even more important than Albert's climatic observations are those left by William Merle,[2] for to him we owe the first detailed systematic weather-record of any duration. His treatise is headed *Consideraciones temperiei pro 7 annis* 1337–1344.[3] It is a monthly record of the past weather. With the year 1340 the details became fuller. It is the abnormalities that attract his attention most commonly, e.g. warm weather in January, storms, snow, ice, &c. The keeping of this record appears to have suggested the composition of a work on meteorological phenomena. Merle's object, as he states at the outset, is twofold, to treat of judgements of future weather and to supply the material requisite for such judgements. The work places very little reliance on popular superstitions in the matter of the signs which it accepts for weather prognostication, and in its attention to past conditions as providing the basis of coming conditions it is more closely aligned to modern than to medieval times.[4]

[1] *Vide De Natura Locorum*: I, 2 and II, 1; also *Libri Meteorum*, II, 1, xvii.

[2] A little-known writer of the fourteenth century. He describes himself, in the title of his work, as 'socium domus de Merton'.

[3] *Vide* Facsimile and English translation by G. J. Symons, 1891.

[4] Lynn Thorndike: *History of Magic and Experimental Science*, Vol. III, p. 144. There is another almost contemporary work on meteorology by

OCEANOGRAPHY

Here, as in other departments of knowledge, the scholar of medieval Christendom was at some pains not to do despite to the orthodox beliefs of the Church. Although the Bible had less to say on this subject than on others, what it did say was regarded as being of paramount importance in view of its context. Thus the statement in the Creation story, that God gathered together the waters in one place,[1] was very generally construed to mean that the waters beneath the firmament, whether in underground reservoirs, oceans, lakes, rivers or in the atmosphere, were interconnecting. No less an authority than St. Augustine had declared that this could only mean that all the waters of the earth came from a single source. This source Isidore termed the 'abyss', maintaining that 'all the waters and torrents return by secret channels to the abyss whence they came'.[2] As this view, moreover, was in line with the observations of the Preacher in the Book of *Ecclesiastes* : 'all the rivers run into the sea, yet the sea is not full : unto the place from whence the rivers came, thither they return again',[3] we are not surprised to find that it was endorsed by medieval scholars. Peter Abelard, Peter Comestor, Hugo of St. Victor, William of Conches, Bartholomew Anglicus and the author of the anonymous *De Imagine Mundi* were among those who subscribed to it. The last-mentioned writer, arguing from analogy, claimed that because blood-vessels penetrated every part of the human body, it was only reasonable to suppose that the earth was riddled with subterranean waterways or channels.[4] Even Frederick II was

Evno of Würzburg : but it is primarily a compilation of rules for predicting the weather, and unlike Merle's work, is not based upon a detailed weather record. See also 'A Meteorological Chronology to A.D. 1450,' by C. E. Britton, *Geo-physical Memoirs* (Meteorological Office, Air Ministry), No. 70. 1937.

[1] *Genesis* i. 9. [2] *Etymologiae :* XIII, 20. [3] i. 7.

[4] 'All in lyke wyse as the blood of a man gooth out and yssueth in somme place, all in lyke wyse renneth the water by the vaynes of therthe and sourdeth and spryngeth out by the fontaynes and welles . . .' (Caxton's translation, *Mirrour of the World*, II, 19.)

assailed with no doubts on this point. 'We know that all waters come from the sea, and passing through divers lands and cavities return to the sea which is the bed and receptacle of all running waters.' [1]

There was no such general agreement concerning the geographical distribution of land and water. Opinion was, in fact, deeply divided between the 'oceanic' and 'continental' theories. The former maintained that the *oikoumenē* is surrounded by water and, accordingly, enjoyed the sanction of the orthodox, in so far as it was believed to be in conformity with Holy Scripture, e.g. '[God] stretched out the earth over the waters.' [2] But there were various views concerning the size of the ocean or oceans which surrounded the habitable earth. Abelard, for instance, said that all the earth's surface except the *oikoumenē* was covered by water.[3] While the author of the fourteenth (?) century *Lumen Animae* held that the sea is a thousand times the extent of habitable regions, and claims that astronomers have proved this, John Calderia of Venice [4] shared Abelard's view—which indeed was a common one—and accounted for the existence of dry land in the northern hemisphere simply by asserting that astrologers believe that while there may be a continual conversion of air into water in the north, yet certain stars near the North Pole hold part of the waters and prevent their submerging the dry land, the residue of them flowing continually to the south.[5] On the other hand, Martianus Capella and Macrobius, both of whom had a large following in such matters, held that there

[1] *Vide* C. H. Haskins: *Studies in Mediaeval Science*, p. 267.

[2] *Psalm* cxxxvi. 6.

[3] *Expositio in Hexaemeron, De Tertia Die*, Migne: *Patrologia Latina*: Vol. CLXXVIII. In commenting on the words 'in one place' Abelard says: 'Id est cedentibus aquis in unam partem terrae, altera detecta est, sicut scriptum est: *Qui fundavit terram super aquas.* Quasi enim aliquis globus ita in aqua constituatur, ut una pars ejus superemineat, ita ille globus terrae in aquis insedit, ut ex una parte eum mare contingeret, et per venas ejus se infunderet, unde nobis fontes vel flumina nascerentur.'

[4] *Fl. c.* 1450.

[5] *Liber Canonum Astrologiae*: fol. 55 *et seq.*

were three areas of land corresponding to our *oikoumenē* in the other quarters of the globe and that, consequently, the ocean could not occupy a very large percentage of the total surface area. Because this view agreed with that advanced in the *Books of Esdras* (which were very popular with medieval readers) it was not branded as heretical. The kind of argument employed for the purpose of rationalizing beliefs of this sort is admirably illustrated by the work of Duarte Pacheco, the Portuguese navigator of the early sixteenth century. Although his *Esmeraldo de Situ Orbis* was composed long after the age of scholasticism, it provides, nevertheless, one of the most palpable evidences of the longevity of clerical authority and also of the manipulation of fresh facts to fit it. In speaking of 'The Size and Greatness of the Earth and Sea and which of them is Greater', Pacheco declares that the earth is not surrounded by the sea as was maintained by many philosophers but 'in its greatness surrounds and contains all the waters in its concavity and centre'—a theory which he proceeds to substantiate from evidence supplied by contemporary explorations. Writing about the year 1505, he says that the Portuguese have discovered

> a very large landmass . . . extending . . . [from] 70° North of the Equator . . . [to] 28½° on the other side of the Equator. . . . Such is its greatness and length that on either side its end has not been seen or known, so that it is certain that it goes round the whole globe.[1]

Now we need hardly remind ourselves that the coastline of the American continent had not been fully explored from Labrador to La Plata by 1505 and that its continuity within these limits was not proved for several years to come. No grounds existed, therefore, for supposing that the Americas spanned the world. This being so, what motive can Pacheco have had in making such an assertion? He himself tells us. He is out to demonstrate that 'the water only occupies a seventh part of the earth, as is shown in the fourth book of the

[1] I, 2. (Hakluyt Society.) Edited by G. H. T. Kimble.

prophet Esdras and the sixth chapter '.[1] For this to be even approximately true it was essential that the new landmass should ' go round the whole globe '. As its extremities were still unknown when he wrote, it was a comparatively simple and safe matter for him to accommodate his facts to his faith.

On at least two oceanographical topics of major importance, namely, tides and the salinity of the sea, the medieval scholar was free to think and write for himself. No part of Scripture could be cited by the protagonists of the orthodox faith. But the ability to think for oneself is at all times hard to come by, and it certainly seems to have eluded the grasp of some of our writers. Thus Alexander Neckam,[2] after quoting the scientific opinions of earlier generations of writers, concludes that the explanation of the ebb and flow of the waters is an insoluble problem, while William the Breton was of the opinion that God alone understands it and that no man can comprehend it either now or ever. ' For us who live our human lot here below, it is sufficient to know the fact ; it is not allowed to us to know the cause.' [3] However, there were more inquisitive spirits who refused to succumb to such intellectual paralysis. These men ranged themselves into two camps, those who supported the *astrological* hypothesis, and those who supported the *physical*. The first, e.g. Robert Grosseteste,[4] attributed tides to the influence of the moon and in so doing enjoyed the support of Posidonius, Pliny and other classical authorities ; the second, e.g. Lambert of St. Omer, attributed them to such terrestrial causes as the impact of ocean currents. There were some, however, who considered neither view to be entirely adequate. Adelard and William of Conches [5] were among these : their solution was to wed

[1] *op. cit.* : p. 13.

[2] *De Natura Rerum* : II, 17 (*Vide* Rolls Series No. 34).

[3] *Gesta Philippi regis Franciae* (edited by H. F. Delaborde) : VI, 550–1. He is, in fact, just as prepared to believe that the tides cause the motion of the moon as *vice versa*, because the sea was created before the moon (*ibid.* : VIII, 73–7).

[4] *De Natura Locorum*. (L. Baur's edition.) p. 69 *et seq.*

[5] *Vide De Philosophia Mundi* : III, 14.

the two theories—with results that were original, if still unsatisfactory. In Adelard's words:

> I admit the recoils (refluxiones) and do not deny the impacts; but I do not admit that they flow together or come into violent conflict. The reason for this I take to be the situation of the earth: for though the channels of the sea are fiercely eager to meet one another and flow together, yet it happens through the interposition of mountains and the somewhat elevated situation of the earth that on their ebb they are drawn from that particular course. The result is that they are held back by their position from the point to which they are impelled by their natural and inherited movement. . . .[1]

Unfortunately, neither William nor any of his supporters explains how this oscillation is set up. The rôle of the moon is fully appreciated by him, but the best he can do to account for the difference between neap and spring tides is to suppose that it is a function of the variations of the moon's power to heat and dry the atmosphere. By arguing that it is at a minimum both when the moon is full and when it is new, he is able to explain the occurrence of spring tides on those occasions, but *why* it should be at a minimum then he is at a loss to tell us.

Giraldus Cambrensis (*c.* 1146–1222) invented an even more ingenious theory,[2] uniting the astrological and the Macrobian, that is, the ocean current, views and the whirlpool theory (derived from Paul the Deacon),[3] which supposed that there were four great vortices at the opposite ends of the earth, towards which 'there is a set current of the waves from all quarters, until, pouring themselves into nature's secret recesses, they are swallowed up, as it were, in the abyss'.[4] Robert of York, or *Perscrutator* as he styled himself,[5] was of the

[1] *Quaestiones Naturales.* (Gollancz's translation, pp. 140–2.)

[2] *Topographia Hiberniae:* II, 3.

[3] *Vide Historia gentis Langobardorum:* I, 6, and F. Nansen: *In Northern Mists:* Vol. I, p. 159.

[4] *Topographia Hiberniae:* II, 14 (Bohn's Antiquarian Library Edition.).

[5] *Fl. c.* 1326.

unusual opinion that the influence of the moon on tides is felt only in the seventh *climata*, i.e. in the northernmost part of the habitable world. Elsewhere the sea does not respond—'resists' is his word—to the moon.[1] He also argued that this theory explains the cause of the whirlpools.[2] Of more permanent value than such imaginative concepts were Giraldus' personal observations of tidal behaviour round the coasts of the British Isles. *Inter alia*, he gave the difference in the hour of high water at such ports as Wicklow, Dublin, Milford Haven and Bristol and established the important fact that the difference is not merely a function of the moon's meridian.[3]

It would be interesting if we could know how Giraldus gathered these *data*. Probably they were pieced together from incidental

[1] *Vide* MS. (untitled) in Cambridge University Library: Ii, i, I, fol. 24r, cols. 1–2. 'Liquet ex predictis occeani tantum partes que latera superfluunt climatis septimi dominio lune obedire; reliquum mare rigidum stare rebelle.'

[2] The reasoning is not easy to follow. Translated literally, the passage concerned reads as follows: 'When the moon approaches the east of the Seventh Climata, then the shores of the sea recede (exeunt) from the east towards the west and from the west towards the east according to the strength of the power of the moon because the influence of the moon in one part of the land (by reason of the fact that the land area is continuous) penetrates (interfluit) quickly, at one and the same moment, into all the remaining parts of the land; and with all the greater force because the other regions of the sea do not respond [to the lunar attraction], while the regions of the earth form an obstruction as is seen when the light of the sun is reflected through an opaque object in an oblique direction. It is obvious, therefore, why, in a similar way, the sea ebbs and flows strongly and simultaneously against its shores in the east and west so that it may find an outlet. In the same way, when the moon is towards the west, the same thing happens as when it is in the east because it attracts the waters so that they cannot flow in accordance with the course of the moon as when it is in the east, because, as it has already been said, the rest of the sea is moved in a contrary manner. The action of the moon is, therefore, turned back (reflectitus) and the sea is agitated by this reflex action [to such an extent] that it rises so much above the earth that it is obvious that its own ebb and flow (reflexio) is not rejected (repudietur). . . . From this it will be seen how whirlpools came to be formed in the ocean, one in the east and the other in the west . . .'

[3] *op. cit.*: II, 1–3.

observations of sailors or fishermen, for certainly no systematic investigation of tidal phenomena could well have been undertaken at Giraldus' time. It is typical of an immense amount of close and accurate knowledge that has always existed along with ignorance and superstition among the more humble workers of the world, knowledge that until recent years has rarely found literary expression.[1]

Another evidence of a practical interest in tidal behaviour is forthcoming in a thirteenth-century tide table, preserved in the British Museum,[2] giving the time of 'fflod at london brigge' for each day of the lunar month and also the hours of moonlight.

The question of the salinity of the sea provided little scope for controversy. Apart from a few naïve conjectures, such as that made by Alexander Neckam[3] and repeated by the author of *Image du Monde* to the effect that the saltness was due to the constant dissolution of saline mountains in the ocean deeps,[4] the general opinion was that of Adelard, viz., that the heat of the sun and the planets was the major cause.

> The real ocean flows through the torrid and centre zone, and it is through the same zone, though indirectly, that the planets have their course: and as a result of the great heat of the stars, the sea itself is necessarily heated and consequently becomes salt. A fact which supports this theory is, that in the sea-board districts near that ocean, sea-water, when dried in the sun on the rocks is, without any artificial process turned into salt.[5]

[1] Wright: *op. cit.*, p. 196.

[2] Cotton MSS., Julius D.7.

[3] *De Naturis Rerum*: II, 1.

[4] Part II, chap. 23. Caxton's rendering of the passage is as follows: 'Ffor in the see ben right grete and hye montaynes and depe valeyes whiche ben ful of biternesses grevous and infected. And the erthe whiche is in the bottom of thise valeyes scumeth for the heete of the sonne upward, whiche medleth with the water in the depe in suche wyse that it draweth the saltnes up by the hete of the sonne, so long til it be medlyd with that other. . . .'

[5] *Quaestiones Naturales*: LI (Gollancz's translation.). Adelard is, for the most part, merely reproducing the ideas of Berechiah, a Jewish writer of the eleventh (?) century.

To these arguments Adelard adds the fact that 'in summer all sea-water is more salt than in winter'.[1] There were various elaborations of this view, needless to say. Michael Scot, for instance, replying to the question of Frederick II on this point, accounts for it thus. 'The waters of the sea are bitter because they are older (than those of the land) and are not exposed to the sun's heat.'[2] Not the least interesting was Gervase of Tilbury's opinion that the amount of evaporation off the equatorial waters would cause them to become extremely viscous and so provide almost as much of a hindrance to navigation as the frozen seas of the North![3]

PHYSIOGRAPHY

During the early centuries of the Middle Ages, text-book expositions of physiographic and geomorphological phenomena were merely reproductions, often very imperfect at that, of arguments employed by the Ancients. The physical views of Pliny and Plato were specially popular—not least, perhaps, because they tended to confirm, in as far as the question of the subterranean character of the earth was concerned, the materialistic conceptions of Hell encouraged by the Church Fathers. Even down to the time of the Aristotelian renaissance men continued to wonder whether the gate to the Abyss was in the volcanoes of the Lipari Isles and Sicily or in the northern isle seen by St. Brandan[4] and at least one thirteenth-century divine declared that he had actually seen the gates of Hell in a lake in Italy.[5]

On most physical topics, however, saner opinions prevailed, particularly after the Dark Ages. The credit for this redounds

[1] *Quaestiones Naturales*: LI. (Gollancz's translation.)

[2] Quoted in C. H. Haskins' *Studies in the History of Mediaeval Science*: p. 296.

[3] *Otia Imperialia*: III, 40. 'Hinc enim, frigore congelatum sub septentrione, fit rigidum: illinc nimio solis ardore spissatum, transeuntibus efficitur ab Euro ad Austrum et meridiem inmeabile.'

[4] e.g. Michael Scot, *vide* C. H. Haskins: *op. cit.*, p. 298.

[5] The Bishop of Pozzuoli, *vide* Gervase of Tilbury: *Otia Imperialia*, III, 103.

mainly to the Arabs, both for their transmission of Aristotle's scientific writings and for their own personal observations. For instance, in a translation of an anonymous Arabic work—perhaps one of Avicenna's—by a certain Alfred of Sareshel, opinions are expressed regarding the origin of mountains and valleys with which modern geologists would have no small sympathy.

> Mountains [we are here told] may arise from two causes, either from uplifting of the ground, such as takes place in earthquakes, or from the effects of running water and wind in hollowing out valleys in soft rocks and leaving the hard rocks prominent, which has been the effective process in the case of most hills. Such changes must have taken long periods of time, and possibly the mountains are now diminishing in size. What proves that water has been the main agent in bringing about these transformations of the surface is the occurrence in many rocks of the impressions of aquatic and other animals. The yellow earth that clothes the surface of the mountains is not of the same origin as the framework of the ground underneath it, but arises from the decay of organic remains, mingled with earthy materials transported by water. Perhaps these materials were originally in the sea which once overspread all the land.[1]

Be it said to the credit of our author that many a worse theory was put forward in the centuries that followed. In one respect, of course (not that we wish to detract from the merit of the argument), the environment of the desert Arabs encouraged speculations of this sort, for the erosive power of wind *and water* is nowhere seen to better advantage than in arid climates.

In the quasi-geographical subject of cosmic cycles [2] and their terrestrial significance, the Arabs developed the old classical notions, but in the course of doing so, they threw out a number of very interesting suggestions on the subject of physical geography. The 'Brothers of Sincerity' of the tenth cen-

[1] *Vide* Sir A. Geikie: *Founders of Geology*, p. 43.

[2] i.e. the number of years taken by a fixed star to revolve round the earth.

tury [1] even postulated changes in the relative position of the lands and seas and so adumbrated the theory of continental drift which was only fully expounded nine centuries later by Wegener. However, in fairness to this later-day scientist it should be explained that the Brothers attributed the movement to the slow changes in the longitudes of fixed stars resulting from the precession of the equinoxes.[2] Almost equally 'modern' was their theory that, in the course of time, cultivated land becomes desert, desert becomes cultivated land, steppes become seas and seas become steppes or mountains.[3]

The views of Christian scholars in these centuries depend less upon personal observation than upon the perusal of Aristotelian literature. Those of Bartholomew Anglicus will serve to exemplify the usually accepted explanation of the origin of mountains. As with the Arab, so with the Englishman, the distinction between tectonic and residual mountains is appreciated, for he tells us that they may be due to 'strong earth-shaking' or to 'the moving of the waters that erode the softer parts of the earth, leaving the harder parts as mountains. The places that are greatly eroded become the basins of seas and the beds of rivers.' His explanation of the earth-tremors is plagiarized from Aristotle's *Meteorologica*.[4] Owing to 'the intermingling of air and fire within the earth, the earth is in some parts hollow and spongy and the winds entering these hollows shake and move parts of the earth, causing earth-

[1] *Vide ante*, chap. iii.

[2] In the fourteenth century, however, the anonymous author of a treatise on Metaphysics and Natural Philosophy (preserved in the Bibl. Nat. Paris, 6,752) suggests that the earth might, in certain parts, be so condensed by the action of water, or cold, as to displace other parts by its greater weight. Such condensation, he submits, would take place so gradually as to be imperceptible. This notion of a slow shifting of the earth's crust and interior he shares with other fourteenth-century thinkers like Albert of Saxony.

[3] But perhaps the inspiration of this theory was 'ancient', for Aristotle says that the seas change their places in the course of centuries.

[4] *Vide Meteorologica:* II, 7, 8.

quakes'[1] which in their turn produce mountains. The connexion suggested here between mountain-building and earthquakes was endorsed by almost every one, but few appear to have linked volcanic activity with it. The majority of scholars, including Hildegard of Bingen, Gervase of Tilbury, Adelard of Bath and Bartholomew Anglicus, were content to accept the classical verdict that volcanoes were caused by vapours and exhalations within the earth's crust, while a minority argued that they were not due to purely physical causes, but were visitations, or portents sent from God. Exactly how the vapours, &c., could give rise to volcanic activity is elucidated for us by Michael Scot. He maintained that, somewhere at the ends of the earth, the winds of the earth's surface were driven into the bowels of the earth where they came into contact with sulphurous, molten rock and so were volatilized and given explosive force. This caused them to be driven back towards the surface. When they reached the vents in the crust they still possessed all the attributes of fire and flame—sparks, ashes and cinders—thereby producing the illusion of actual fire. 'A great mercy of God it is,' he adds, 'that this sulphur and these rocks are hidden away far beneath the surface, for otherwise the winds blowing over them would destroy the world with their heat.'[2]

[1] cf. Isidore: *Etymologiae*: XIV, 1, iii. 'Others have it that the earth is sponge-shaped, and its fallen parts lying in ruins cause all the upper parts to shake. The yawning of the earth also is caused either by the motion of the lower water, or by frequent thunderings, or by winds bursting out of the hollow parts of the earth.' Neckam, Honorius, and the author of the *Image du Monde*, also favour this point of view. In the *De Mundi Constitutione* (of unknown authorship, but formerly attributed to Bede) the writer speaks of those who say that 'the earth contains the animal leviathan, and that he holds his tail after a fashion of his own, so that it is sometimes scorched by the sun, and so the earth is shaken by the motion of his indignation. . . .' *Vide* A. D. White: *The Warfare of Science with Theology*, Vol. I, p. 327.

[2] *Vide* text in C. H. Haskins' *Studies in the History of Mediaeval Science*, p. 297. 'Sed cum misericordia sit maxima in dispositione constitutionis mundi, hunc sulphurem et hos lapides locavit inter terram propter melius, nolens quod mundus taliter destruatur, unde voluntate Dei flamme dictorum locorum nec mundum destruunt nec loca sit propinqua . . .'

Although they did not excite so much attention (possibly because they were less spectacular than volcanoes) thermal phenomena are by no means ignored by medieval students of science. But it was not until the thirteenth century that the geysers of Iceland became generally known to Europeans. Saxo Grammaticus, the Danish historian of that century, describes them soberly and without comment in the following words. 'There are many . . . wells and becks also which are sometimes flooded with water, shooting it out high into the air and covering everything near with foam.' [1] In Italy, where the behaviour of hot springs could be watched in more congenial surroundings, controversy was more alive. The general view, dating from the days of Seneca, was that they were caused by subterranean waters flowing over veins of sulphur. Michael Scot put it thus:

Waters were created with inexhaustible virtue of pouring forth so long as the world endures, and they move about in the earth like blood in the veins, the quality of the water depending on the earth through which it passes, and its heat coming from dry, hot rocks, especially sulphur.[2]

In the fourteenth century this view was opposed by a certain John de Dondis [3] who maintained that they were heated by subterranean fires and gases. The 'hot gases are hot exhalations generated and elevated by virtue of the sun and stars', and he justifies this astrological argument by reminding his readers that Aristotle in the first book of his *Meteorologica* declares the elements to be 'led, ruled and governed by the stars'.[4] But de Dondis is disinclined to attach too much importance to explicable relationships.

I have learned from long experience [he declares] that there is nothing which is not marvellous and that the saying of Aristotle is true . . . that in every natural phenomenon there is something

[1] *Gesta Danorum, Praefacio* (Holder's edition): p. 6.

[2] Quoted in C. H. Haskins' *Studies in the History of Mediaeval Science*: p. 296.

[3] *Fl. c.* 1380.

[4] *Vide De balneis* (1553 Edition): fols. 94–108.

wonderful, nay, in truth, many wonders. . . . We are born and placed among wonders and surrounded by them, so that to whatever object the eye first turns, the same is wonderful and full of wonders, if only we will examine it for a while.[1]

Thus does de Dondis (and he speaks for his century) mingle his credulity with its scepticism.

The superficial processes of geology attracted very little attention in the Middle Ages. Commonplace and stable elements of the earth's surface such as hills, valleys and plains were more or less taken for granted by those who sought to explain the secrets of Nature. A plain was a plain and a valley was a valley because God made it so. It was the eccentricities rather than the normalities of behaviour that were usually regarded as worthy of record. The ordinary work of the mechanical and chemical agents of erosion consequently finds no place in the text-books of our period. But rivers with underground courses; rivers flowing in opposite directions from the same source (as some were thought to do); rivers with unusual regimens; rivers with tidal bores; rivers which change their course; springs[2] and fountains[3]—especially if they were reputed to have magical properties; lakes—especially if they had famous associations, e.g. the Dead Sea; and glaciers: such features are quite commonly described and discussed.

The usual medieval belief about the flow of rivers was that it could not be explained as the mere result of precipitation, but that it was supplied from the sea and that the water found on mountain-tops came up through pores in the earth by capillary attraction. The belief, like most others, did not originate with the Middle Ages. In all probability it found its genesis in the idea cited by Aristotle, but regarded by him as

[1] *Ibid.*: fol. 95v.

[2] e.g. the spring in Munster, Ireland, which when approached by any human being, always proceeded to inundate the entire province with rain. *Vide* Giraldus Cambrensis: *Topographia Hiberniae*, II, 9.

[3] e.g. the Fountain of Youth described in the first letter purporting to have come from Prester John.

erroneous,[1] that 'the size of the subterranean cavities is what makes some rivers perennial and others not'. The view is elaborated admirably by Ristoro of Arezzo, a late thirteenth-century writer.[2] In common with his contemporaries, Ristoro accepted the Aristotelian contention that as earth was the heaviest of the four elements, the sphere of water should, theoretically, cover the whole earth, just as it in turn was surrounded by the lighter elements of air and fire. At the same time, he argued that a portion of the earth's surface had to be uncovered by water in order that the heavenly bodies might exert their generative force upon the earth. This portion, Ristoro maintained, amounted to one quarter.[3] The waters displaced from this portion of dry land were piled up elsewhere, and owing to the virtue of the heavenly bodies defied the predisposition of water to find its own level. If once the stars relaxed their hold on these reservoirs, then naturally a universal deluge would result. The downward pressure of these reservoirs forced water up through the spongy, porous crust of the earth, even to the head-waters of mountain rivers.[4] This view, blind as it was to the very obvious correlation between the run-off of precipitation and river volume, continued to be supported right down to the closing years of the fifteenth century.[5]

[1] *Vide Meteorologica :* I, 14.

[2] *Vide La Composizione del Mondi di Ristoro d'Arezzo :* VI, 5, 7.

[3] *Vide infra*, chap. ix, p. 217. Cf. Robert of York (*Perscrutator*), *op. cit.* 'Est enim ut prius patet habitatio hominum quarta pars terre et eius figura est similis medietati circuli.'

[4] Cf. Adelard : *Quaestiones Naturales*, LVI. 'Since in the bowels of the earth rivers divide their courses in many ways, it may happen that a stream may enter ground which is blocked by obstacles on all sides, and is therefore compelled to flow out upwards, if such an exit exists : continually rising, it is bound to come to the surface—a thing which you can see exemplified even in works built by man for holding water' (Gollancz's translation.).

[5] *Vide* John Michael Albert of Carrara : *De Constitutione Mundi*, XI, 5, 7. The same theory was still accepted here and there in the seventeenth century. *Vide* Lynn Thorndike : *Science and Thought in the 15th Century*, p. 203.

As with the Ancients, the hydrographical problem to receive most attention was the flooding of the Nile. Even so, the Middle Ages gave birth to no new theories; rather did it revert to the palpable explanation attributed to Thales that the flooding was caused by the sand bars which accumulated at the mouth of the river during the summer, when the Etesian winds blow—a view popularized by Isidore in his *Etymologiae*. The more plausible hypothesis of Anaxagoras—accepted by Pliny and Strabo—that the phenomenon was caused by the melting of snows on the Ethiopian mountains—found little favour in this era. At the very end of the medieval period we even find Duarte Pacheco rejecting all the classical theories and substituting in their place the view that the floods are the result of rains in the southern Hemisphere. Pacheco, obsessed with the importance of the discovery of the Cape route to India, suggests that they are due to summer rains in the Cape region. In his estimation it only takes 'a few days' for the effects of them to be felt in Egypt, 'owing to the rapid flow of the river'.[1] What sort of gradient the Nile would need in order to traverse the 8,000-odd miles separating the Karroo from Cairo in so short a time, is left to our imagination. That the real reason for the floods was known and appreciated in those days is apparent from Benjamin of Tudela's[2] *Itinerary*. Here we read that

persons who inquire at the reason of the rise of the Nile, are told by the Egyptians that it is caused by the heavy rains which fall in the Country of Abyssinia, the Havilah of Scripture, and which is elevated above the level of Egypt: this forces the river out of its bed and inundates the whole country'.[3]

Leo Africanus, writing at the very end of our period (*c.* 1500), could also have told Benjamin how long it takes the flood waters to reach lower Egypt, namely, about forty days.[4]

[1] *Esmeraldo de Situ Orbis*, *op. cit.*, I, 4.

[2] *Fl. c.* 1160.

[3] *The Itinerary of Benjamin of Tudela*, translated and edited by A. Asher: p. 151.

[4] *History and Description of Africa* (Hakluyt Society): Vol. III, p. 936.

THE STUDY OF ENVIRONMENTAL INFLUENCES

While it is true that geography considered as the study of action and reaction between man and his environment is a product of comparatively recent times, observations upon this subject go far back into the past. As far as the Middle Ages are concerned, the literature on this aspect of geographical science consists of scattered fragments, all of them of fairly late date. The cloistered life of the early Fathers provided few opportunities for the contemplation of the cultural landscape. It was not until men began to travel as they did in the Crusading epoch that they began to perceive the varying relationships existing between landscape and life. If we are tempted to complain that their observations savour rather strongly of 'determinism', we need to ask how it could very well be otherwise seeing that medieval man was essentially the creature of his circumstances. Further, he enjoyed only very limited opportunity for first-hand study of the matters involved and for the collation of relevant material; consequently his generalizations were usually questionable, if not palpably erroneous. Bernard Sylvester, for instance, assures us that 'fruitful land gives birth to wolves, desert to lions, arid lands to serpents, woods to bees'.[1] Hildegard of Bingen avers that light sandy soil is retentive of moisture in the form of large drops, and because of this quality favours the growth of vines and apples, though rather less well adapted for the raising of grains. The latter may better be cultivated upon soil characterized by finer or more minute water particles. Black soil, on the other hand, is too cold and damp to produce more than modest yields of anything. Reddish soils, she maintains, preserve a good moisture balance and hence are highly productive.[2] Unfortunately for her practically-minded readers, this classification scarcely tallies with the facts.

[1] *De Mundi Universitate*: p. 21 (edited by C. S. Barach and J. Wrobel).
'Cepit enim fructuosa, deserta leones,
Arida serpentes, pars nemoralis apros.'

[2] *Subtilitates diversarum naturarum creaturarum*: I, 9, in Migne: *Patrologia Latina*, Vol. CXCVII, col. 1214.

Some of the most valuable and interesting observations to be made upon the subject deal with the influence of climate upon plants and man. Thus William Merle, in his work on meteorology, devotes a whole chapter (one of twelve) to the climatic causes, precedent and concomitant, of scarcity of crops, especially in certain parts of England.[1] Among other things ten ways are listed in which humidity may produce such a failure. And Giraldus Cambrensis, not content with the ordinary statements in circulation about the climates of the *oikoumenē*, advances the view that the West with its stormy, damp weather is more healthy than the East.

> In those [oriental] countries . . . the heavens terrify you with their thunders, and flash their lightnings in your eyes. The blazing sun allows you no rest. If you eat too much, death is at the gate; if you drink wine undiluted with water, death is at the gate. Besides this, poison threatens on all hands. . . . We (in Ireland) sleep secure in the open air . . . we fear no wind piercing us with cold, prostrating our strength with heat or carrying pestilence in its blast. The air we breathe and with which we are surrounded, lends us its beneficent and salutary support.[2]

Moreover, he gives it as his considered opinion that Ireland possesses the best climate in the whole world[3]—an opinion which may not be endorsed by the inhabitants, but which, nonetheless, agrees remarkably with some twentieth-century views. It was even said that in Ireland men grow so old that they ask to be taken to some less salubrious climate in order to be able to die at last![4] In company with Bartholomew Anglicus and Gervase of Tilbury, Giraldus goes still further and re-affirms the view common among classical geographers that the character of different peoples varies with the different climatic conditions.

> The nearer . . . we go to the regions of the East and warmer

[1] *Vide ante*, p. 160.

[2] *Topographia Hiberniae*: I, 26–7.

[3] *Ibid.*: I, 25.

[4] *Vide* Pompilius Azalus—*alias* G. da Fontana: *Liber de omnibus rebus naturalibus quae continentur in mundo* . . . *c.* 1450, fols. 117v, 140v.

climates, . . . the people, thanks to a brighter atmosphere, although slender in person, are of a more subtle intellect. Hence they have recourse to poison rather than to violence for success in their schemes and gain their purposes more by their arts than by their arms. But when we come to the western parts of the world, we find the soil more sterile (*sic*), the air more salubrious, and the people less acute, but more robust: for where the atmosphere is heavy, the fields are less fertile than the wits.[1]

Bartholomew endeavours to explain how these presumed distinctions may come about.[2] 'The north wind dries and cools the land, yet owing to its clearness it is clear and subtle'; because of its coolness the pores of the body are closed and its heat retained. For this reason 'men in the north are tall in stature and fair in body'. The south wind, on the other hand, is hot and moist. 'Therefore the men of the southern lands are different from those of the north in stature and appearance. They are not so bold nor so wrathful, nor so angry . . .' But opinions were divided even here. John Calderia (a fifteenth-century Venetian physician), for instance, employing stock astrological arguments, averred that the differences between the south and the north are due to the fact that the influence of the stars was weaker in the southern lands than in the north and that the signs of the Zodiac were less fortunate. 'Therefore the South is of necessity sterile, morbid and pestiferous; the north fruitful, jocund and most healthful.' This contention he proceeds to strengthen by an argument based upon the nature of the South wind. Because of its humidity it induces corruption of matter and mind. Men of the South, therefore, are 'savage, brutal, without religion and often cannibals. To make matters worse there is a multitude of spirits and demons there.'[3] Al-Mas'ūdī essays to account for

[1] *Topographia Hiberniae*: I, 27. Cf. Gervase: *Otia Imperialia*, II, 3. 'According to the diversities of the air the Romans are grave, the Greeks fickle, the Africans sly, the Gauls fierce, the English of great ability (*ingenio potentiores*) and the Teutons robust.'

[2] *De Proprietatibus Rerum*: XI, 3.

[3] *Liber Canonum Astrologiae*: fol. 58r *et seq.*

the abnormalities of the negro by arguing that the heat of the equatorial belt

> predisposes in the humours a tendency to go to the upper parts of the body: hence their eyes are large, the lips thick, the nose flat and big and the head high. The crasis of the brain is, therefore, out of proportion and the mind cannot perfectly manifest its action; the nicety of distinctions and the action of the understanding are confused.[1]

Occasionally, the medieval student of geographical influences displays an altogether more intelligent understanding of the interplay of cause and effect. Thus Giraldus tells us in his description of Wales[2] that

> as the southern part of Wales near Cardiganshire, but particularly Pembrokeshire is much pleasanter on account of its plains and sea-coast, so north Wales is better defended by nature and more productive of men distinguished for bodily strength;

further, that the natural defences of the north are such (e.g. the steepness of its hills and valleys, the size of its forests, swamps and lakes) that it presented a much tougher proposition to any invading army than the wide plains of the southern counties. Gervase of Tilbury also remarks[3] upon the same thing, specifying how, on sighting an enemy, the Welsh take to the bog-lands on the mountains which they can cross with ease and agility resulting from long acquaintance. Bartholomew is another to make allusions to the broad correlations between land and life. The fertility of the soil of France is a function of its maritime location and its many noble rivers;[4] the mineral and agricultural resources of England are a big factor in making that country 'the plenteousest corner of the world, so rich a land that it needeth no help of any land, though every other land needeth help of England'.[5] The coldness of the climate and the scarcity of suitable plough land are held responsible for the conspicuous absence of cereals in Iceland.[6]

[1] *Meadows of Gold. . . .* Translated by A. Sprenger; p. 181.
[2] *Descriptio Kambriae:* I, 6 *et seq.* [3] *Otia Imperialia:* III, 82.
[4] *De Proprietatibus Rerum:* XV, 57. [5] *Ibid.:* XV, 14.
[6] *Ibid.:* XV, 174.

Such correlations, inadequate as they may be from any modern view-point, are yet a considerable advance upon the magic-ridden beliefs current at the time.

Among the Arabs, the influences of climate were well understood. Ibn Khaldūn, one of their greatest philosophers, even formulated a theory of historical development which took due cognizance of the facts of climate and topography as well as of the moral and spiritual forces at work. In his endeavours to investigate the laws of national progress and decay, Ibn Khaldūn may be considered to have discovered—as he himself claimed [1]—the true scope and nature of geographical inquiry.

[1] *Vide* P. K. Hitti: *History of the Arabs*, p. 568.

CHAPTER VIII

MAPS IN THE MIDDLE AGES

AS we conceive it to-day, the primary function of a map is to provide the reader with a graphical record—climatic, topographical, economic, as the case may be—as accurate as the limitations of scale and human knowledge will allow. But its function was not always thus construed. In the Middle Ages world-maps and maps in general ordinarily filled any rôle but that, for while they reflected current ideas, including the quasi-scientific theories of the Greeks, pagan mythologies and Christian systems of cosmography, very few of them—almost none before A.D. 1400—reflected the real extent of contemporary knowledge. Yet great as was the prepossession of medieval authors with classical and clerical erudition, it can hardly be held responsible for the inaccuracy of the average world-maps. In them lands well known and well described in the text-books of the time are distorted and exaggerated beyond recognition.

Notwithstanding, these maps are of absorbing interest and value, because they preserve the materials upon which the ordinary student of the time based his conception of the world around him. 'We can catch much more of the flavour of the popular geography of the Middle Ages by a hasty glance at one of the crude Beatus representations of the world than by ploughing through many of the dry pages of compilations like *De Imagine Mundi*.'[1] Such works, it is readily conceded, had a great vogue, if only to judge from the frequency with which they were listed in library catalogues, but their chief use was as source-books of 'popular' knowledge, while the maps con-

[1] J. K. Wright: *Geographical Lore of the Time of the Crusades*, p. 247.

tained in them were constantly being copied and modified for the purpose of illustrating books on theology and philosophy. Since pictures make a readier and, oftentimes, a more abiding impression on the understanding than the written word, it is not difficult to see why the geographical ideas of the medieval student were materially influenced by these maps.

One of the most striking illustrations of their popularity and, incidentally, of the love of ornamentation in the Middle Ages, is provided by the extraordinary varied occurrence of the *mappaemundi.* Some accompany, appropriately, text-books on history and cosmography, as for example, those found in the *Polychronicon* of Ranulf Higden[1] and the *Tractatus de Imagine Mundi* of Pierre d'Ailly.[2] Some appear never to have been attached to a written work, but 'made to order'. In this category fall the Catalan Atlas of A.D. 1375, the Este World Map *c.* 1450,[3] and the Fra Mauro Planisphere of 1459.[4] Others take the form of miniatures ornamenting the capitals of illuminated manuscripts, e.g. the map inscribed in the letter P of an early fifteenth-century manuscript of Pomponius Mela's *De Situ Orbis*, while at least one adorns the reverse of a medal and another the exterior of a sacramental vase. In the aggregate, it is probably true to say that the great majority of these *mappaemundi* are to be regarded as works of art and not of information. Their authors were creating something very different from the modern cartographic sheet that stands on its own merits as an essentially utilitarian document, scientific in construction. They would have branded any man a fool who might have supposed that he could determine the distance from London to Jerusalem by putting a ruler across a map. For most of them a map was a somewhat elastic framework within which subjects of popular, rather than of scientific, interest could be delineated. The taste of the age ran, as we have already had occasion to notice, in favour of the marvel-

[1] *Fl.* mid-fourteenth century; *d.* 1364. [2] *Vide* Plate No. XIV.

[3] *Vide* G. H. T. Kimble: *Memoir on the Catalan World Map of the R. Biblioteca Estense at Modena*, p. 5, and Plate No. IX.

[4] *Vide* Plate No. XII.

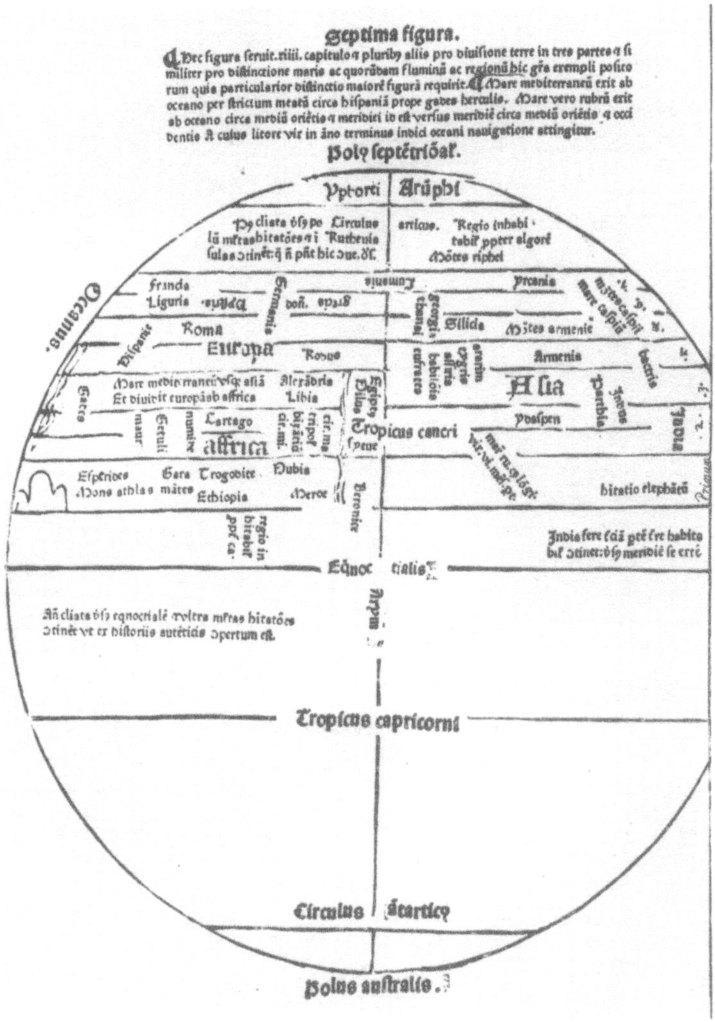

XIV

DIAGRAM OF ZONES FROM PIERRE D'AILLY'S *Imago Mundi*

(?1480)

lous, and the medieval world-map was in no small measure devoted to the elaboration of this department of literature.

Let anyone compare the Hereford Map with the Romance of King Alexander [1] and he cannot fail to see the close resemblance in the spirit and even in the features of the two documents. In short, a mediaeval *mappamundi*, to be duly appreciated, must, to a considerable extent, be regarded as an illustrated romance.[2]

All the same, this is not to deny practical value to the *mappaemundi*. The Este World Map, though embellished with castellated towns, ships and portraits of African princes, attempts both to furnish an up-to-date picture of the world and to resolve the ancient riddle of *Africa nondum cognita*. The Borgian World Map [3] of about the same date, i.e. *c.* 1450, sets out to achieve something even more practical, for, according to A. E. Nordenskiold, it was most probably composed 'to illustrate instruction in the elements of the globe, or, more correctly, in the geography, the natural conditions and ethnography of the earth's disc.' [4] Even more noteworthy, and in this respect it is almost unique among medieval maps, is the fact that it seems to have been drawn, not by some scholar by the study of older authorities, more or less classical, but by a much-travelled and observant man from what he had seen and heard. Even he, however, could not resist the temptation to tickle the palate of his readers, for he fills the empty, unexplored continental spaces with all manner of legendary and traditional characters, including the Magi of the Gospel story.

In view of the strength of the religious bias throughout this era, the use of biblical material and the pursuit of biblical objectives are, of course, not surprising. Some maps were executed to portray the spread of the Christian faith over the earth. This *motif* is clearly expressed in one of the 'Beatus'

[1] For a discussion of the medieval legends relating to Alexander, *vide* Lynn Thorndike : *History of Magic and Experimental Science*, Vol. I, p. 551 *et seq.*

[2] W. L. Bevan and H. W. Phillot : *Mediaeval Geography, an Essay in illustration of the Hereford mappamundi*, p. xxii.

[3] *Vide* Plate No. VIII.

[4] *An Account of a Copy of a 15th Century Map of the World* : pp. 14–15.

maps [1]—the Osma copy (A.D. 1203)—in which we are shown a series of pictures of the Twelve Apostles, each in the locality where tradition placed his preaching and diocese. Others, such as the 'Cotton' map, had for one of their main objects the location of the Twelve Tribes of Israel.[2]

The clerical hold on scholarship was responsible for two of the most conspicuous features of the typical world-map; firstly, the prominence given to biblical topics and topography and, secondly, the survival of certain traditions at a time when fresh knowledge was making them untenable or at least demanding their modification. The Terrestrial Paradise, for instance, forms an almost constant component of the *mappa-mundi*, and what could be more natural? No orthodox Christian in the Middle Ages doubted the existence of this original home of mankind as a fact of contemporary history. Many writers devote long chapters to the description of its delights, though none from first-hand enjoyment of them! Even Mandeville, the most romantic geographer of the age, confesses that he had not visited it on account of his unworthiness, but that he had derived his information about it from trustworthy men. John of Hesse (Hese) professes to have seen it from a distance in the Far East,[3] and John Marignolli was assured by the natives of Ceylon that Adam's Peak was only 40 miles distant from Paradise and that on a good day it was possible to hear the water falling from the river which 'went out of Eden to water the Garden'. Typical of the 'circum-

[1] Named after an eighth-century Spanish priest who wrote an illustrated *Commentary on the Apocalypse*. Among the illustrations was a *mappamundi* which was copied, or rather modified with every transcription of the work. About ten copies of the map are known. *Vide* Add. MSS. 11,695 (British Museum).

[2] Cotton MSS., Tib. B.v, fol. 58 (Brit. Mus.). *Vide* also Henry of Mainz's map accompanying his transcript of the anon. (Honorius of Autun's?) *Imago Mundi* (Corpus Christi College Library MS. LXVI, Cambridge).

[3] *Vide Peregrinatio Joannis Hesei*. He (*fl. c.* 1389) also assigns a terrestrial position to Purgatory, possibly on the authority of Dante who tell us that the Earthly Paradise was situated in the Southern Hemisphere on the summit of the mount of Purgatory—antipodal to Jerusalem. *Vide* Appendix, pp. 241–4.

stantial' descriptions of this earthly Eden are those coming from the pens of Gervase of Tilbury [1] and Ranulf Higden,[2] who based their statements mainly on the opinions of the early Fathers, Augustine, Basil and Ambrose. But the authority upon whom the map-makers relied mostly was Isidore, whose statement that Paradise was 'hedged about on all sides by a long wall of flame . . . in such a way that the fire reached almost to the sky',[3] is vividly portrayed in the Hereford map. The vitality of the tradition was so great that this 'Garden of Delights', with its four westward flowing rivers, was still being located in the Far East long after the travels of Odoric and the Polos had demonstrated the impossibility of any such hydrographical anomaly,[4] and the moral difficulties in the way of the identification of Cathay with Paradise. Thus Andrea Bianco (Bianchi) in his World Map of *c.* 1436 shows 'Paradiso Terrestre' adjoining C. Comorin, whilst four rivers are exhibited as flowing through the centre of India—one into the north of the Caspian, near Agrican, that is Astrakan (the Volga), a second into the south of the Caspian, near Jilan (Araxes ?), a third into the Gulf of Scanderoon (Orontes ?), while the fourth is the Euphrates.[5]

Other Old Testament stories to be commemorated were the fortunes of Noah's Ark, the punishment of Lot's wife, the destruction of Sodom and Gomorrah, the sojourn in Egypt and the Exodus.[6] However, of all the topics to be given pictorial expression, few enjoyed a wider vogue than those concerning the lands of Gog and Magog. The reason is not far to seek.

[1] *Otia Imperialia:* I, 10, iii. [2] *op. cit.*

[3] *Etymologiae:* XIV, 3, iii.

[4] The embarrassment arising from the knowledge that the sources of the rivers were mutually remote was banished by assuming that each of the streams, on leaving Paradise, went underground and reappeared at their respective sources.

[5] The physical existence of the Earthly Eden was believed by many people, long after the Middle Ages. Its location was still an academic issue when Bishop Huet of Avranches wrote his *Tractatus de Situ Paradisi Terrestrii* in the eighteenth century.

[6] e.g. Higden, Psalter and Hereford maps. *Vide* Plate No. XV.

Throughout the Middle Ages it was generally believed that the inhabitants of these lands were so powerful that the only thing which restrained them from overwhelming other nations was an immense barrier [1] thrown up by Alexander the Great, who had himself driven them back to their home in the bleak plains bordering the Northern Ocean. From here they would come forth in 'the last times' to carry death and destruction throughout the fair lands of Christendom. In the thirteenth century the anticipation of this dreaded event overshadowed the counsels of thinking men to such a degree that even so enlightened a theologian and scholar as Roger Bacon recommended the study of cosmography in order to ascertain the most likely time for this invasion to occur.[2] The prophecies of *Ezekiel*,[3] coupled with a statement in the Book of the *Revelation*,[4] supplied the Scriptural warrant for the belief.

The same excessive regard for the letter of prophetic utterances in the Old and New Testaments also explains the occupation by Jerusalem (frequently down to the fifteenth century) of an orbo-centric position in world-maps.[5] This is particularly noticeable in the cases of the thirteenth-century Psalter Map [6] and of Higden's fourteenth-century map. Once again the Book of *Ezekiel* supplies us with the explanation, for there we read: 'I have set it in the midst of the nations and the countries that are around about her.' [7] Much the same reason may be held to explain the eastern orientation of most medieval maps—it was not until Matthew Paris produced his maps of England that the north came into its own—and the representation of biblical places, irrespective of their medieval importance. Thus the above-mentioned maps give special prominence to

[1] A confused reference to the Great Wall of China?

[2] *Opus Majus*, *op. cit.*, pp. 320–323. [3] xxxix, 2. [4] xx, 8.

[5] Even as late as 1664 the eminent French priest Eugene Roger in writing of Palestine, dwelt on the references in the Old Testament in order to prove that the exact centre of the earth is a spot marked on the pavement of the Church of the Holy Sepulchre in Jerusalem. *Vide La Terre Sainte*: pp. 89 *et seq*.

[6] Add. MSS. 28,681, fol. 9 (British Museum). *Vide* Plate No. XV.

[7] v, 5.

XV

THE PSALTER MAP—THIRTEENTH CENTURY

(*Brit. Mus., Add. MSS.* 28,681)

ruined towns such as Nineveh, Babylon, Jericho and Chorazim, without so much as referring to the Crusading ports of Jaffa and Acre. In order to accommodate all this information, Palestine and the adjacent Bible lands are given a territorial extent out of all proportion to their size. In the Psalter Map they occupy more than a third of Asia and in the Higden Map scarcely less.

Rather surprisingly, in the light of this emphasis, hardly any maps in these centuries take a literal view of those passages of Scripture regarded by the 'extremists' as indicating a rectangular earth. Whereas many a cosmography in the Dark Ages—Cosmas' *Christian Topography* for example—was inspired by such phrases as 'the four corners of the earth', these statements do not appear to have convinced the cartographers. Santarem, however, claims [1] to see traces of their influence in one or two *mappaemundi*, notably in the Mela manuscript map of 1417. It is certainly contained within a square—the upper part of the initial letter P—but the author has inscribed a *circular* earth within this framework. The angels trumpeting in the four otherwise unoccupied corners confirm Santarem in his view, for he sees in them the four angels of *St. Matthew's Gospel* gathering up the elect from the ends of the earth.

In the course of centuries the *mappaemundi* became more and more removed from reality, for so great was the stranglehold of tradition, classical as well as ecclesiastical, upon the medieval mind that rather than accept facts uncongenial to their way of thinking, map-makers not uncommonly preferred to ignore them and in their stead employ imaginative and schematic designs. On these it would frequently be quite difficult to make out which continent is which, were they not named.[2] The Ptolemaic world-map, meantime, remained in total eclipse, save for the brief period when it attracted the attention

[1] *Essai sur l'histoire de la cosmographie pendant le moyen âge:* Vol. I, p. 244.

[2] e.g. Reims *mappamundi* (P. Mela MS.), *c.* 1471 and the anon. fourteenth-century MS. No. 4,126 (Latin Series) Nat. Bibl. Paris.

of the Arabs. Idrīsī's map (*c.* 1154) is the most notable—indeed, we might almost say the only—medieval map, prior to the early fifteenth century, in the Ptolemaic tradition. But this is not to say that classical influence was non-existent in the cartography of the Middle Ages. On the contrary, several world-maps present, albeit in somewhat misshapen guise, the theories of the early Greeks, such as Homer and Anaximander, for they show the figure of the earth, on a flat projectionless surface, in the form of a disk, surrounded on all sides by the ' ocean river ', a view that was current down to the time of Pliny.[1] Classical influence is further manifested in the use made of the creatures of Greek and Roman mythology popularized by Hesiod, Homer, Pliny and other fable-mongers of Antiquity. Such were the *Sciapodae*—a race of men whose feet were large enough to be used as sunshades !—the *Monoculi* and the *Cynocephalae*.[2] As belief in the existence of human monstrosities was justified by no less an authority than St. Augustine,[3] it was almost universally adopted in medieval Christendom. In large measure this explains the popularity of the *Mirabilia* and the *Bestiaries*.[4] A third and last manifestation of classical influence is supplied by the map nomenclature. Places famous in the ancient world live again in the *mappaemundi*. Troy and Carthage vie in importance with Rome and Jerusalem.[5]

With this pronounced predilection for classical subjects goes a rather unexpected ignorance of ancient topography. In the Hereford Map, for example, Delphi is confounded with Delos, Thermopylae becomes an inland range and Corinth stands remote from any suspicion of an isthmus. Then again Patmos is transported to the Black Sea, and Gades (Cadiz) is represented as a large island in the middle of the Straits of Gibraltar. Instances of this really culpable confusion could be multiplied, but they would do little more than underline the

[1] e.g. Marino Sanuto World Map, *c.* 1320. *Vide* Plate No. XIII.
[2] e.g. Walsperger World Map, 1448. *Vide* Plate No. XVI.
[3] *De Civitate Dei :* XVI, 8.
[4] *Vide ante*, chap. iv.
[5] e.g. Hereford and Ebstorf World Maps, *vide* Plate No. VI.

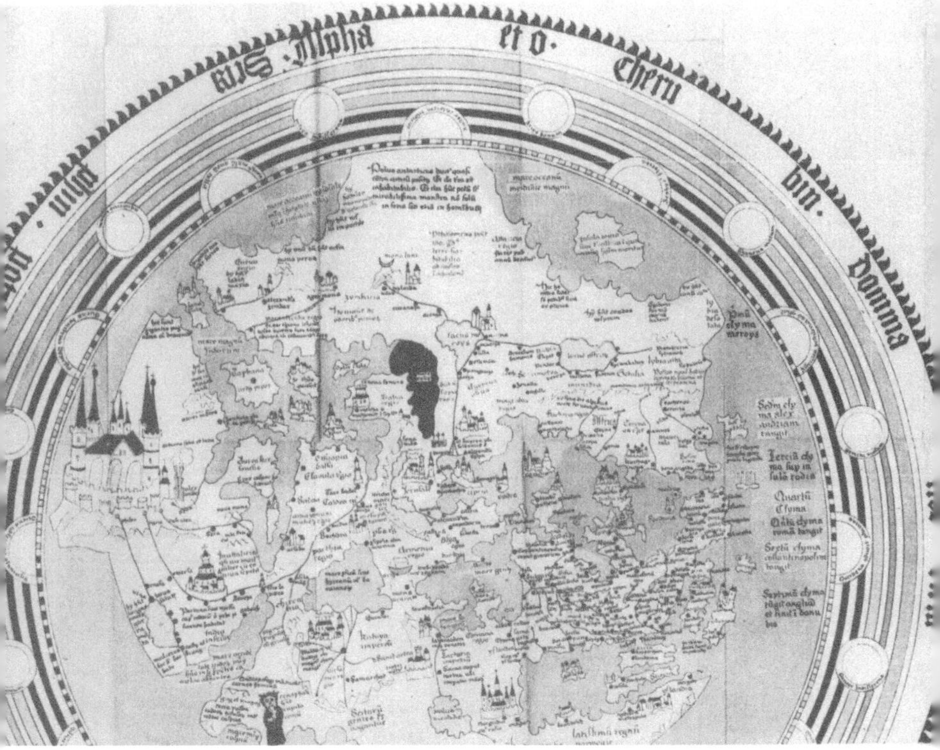

XVI. THE WORLD—ACCORDING TO ANDREAS WALSPERGER (1448)
(Biblioteca Vaticana: Rome)

essential truth of Santarem's dictum that the maps of the age are 'the barbarous offspring of those of the ancients'.[1]

There are, however, at least four maps of this pre-Portolan period which deserve a more generous criticism. They are the Matthew Paris maps of Great Britain.[2] It is true that opinions differ concerning their cartographic value, but it is important to realize that they were designed for a particular purpose—namely to indicate the main route from Dover to St. Albans and thence to Durham. The showing of this as a straight line naturally results in a considerable distortion of the coast-line. When allowance has been made for this, it is apparent that Matthew Paris possessed no small knowledge of the geography of the country. In all, more than a hundred places of importance are portrayed, inland towns outnumbering seaports—in sharp contrast to the Portolan charts of the next century. Conspicuous among these towns are the centres of Benedictine rule, familiarity with which Paris, who belonged to that order, no doubt gained from his attendance at its provincial convocations. Not the least significant feature is the statement on one of the maps that 'Anglia' is 800 miles long and 300 miles in breadth from Dover to St. Davids—a statement which provides us with an approximation to a scale.[3] Improvements as they are upon their 'barbarous' contemporaries, they are far from being practical maps, and along with the former, serve to show the need of a real map science.

Naturally this need was long standing among seafaring people, though exactly when it first came to be met we have little opportunity of discovering. The outburst of maritime enterprise in the twelfth and following centuries would certainly seem to have had something to do with the fulfilment of the need. This came in the form of the sailing chart, or

[1] *op. cit.*: Vol. I, p. 177. On medieval world-maps in general, Konrad Miller's *Mappaemundi, die ältesten weltkarten* (1895–8), is still a standard work.

[2] *Vide Four Maps of Great Britain, designed by Matthew Paris, c. 1250.* Published by the British Museum, 1928.

[3] For a full analysis of these maps, *vide* J. B. Mitchell: 'Early Maps of Great Britain' in the *Geographical Journal*, Vol. LXXXI, pp. 27 *et seq.*

Portolan chart, to use the name by which it was known in the region of its early development, viz. the Mediterranean lands. Although the oldest extant specimens of these maps date only from *c.* A.D. 1300, it is quite plain from the excellence of their technique that their origins are to be sought much further back. Probably many years, even centuries, of painfully bought experience went to create them; the notes, sketches and oral traditions of generations of captains and pilots are without doubt to be seen in the results. As common needs not uncommonly beget kindred results contemporaneously, and even in widely separated regions, it is not improbable that such charts began to appear in different parts of the world round about the same time. Arab pilots guiding their vessels across the Indian Ocean must have sensed the need quite as strongly as the Genoese traders in the more sheltered arms of the Mediterranean. The supposition that there may have been Arabic charts in earlier centuries is not entirely without foundation, for Marco Polo confesses that he drew his knowledge of the coast of Ceylon from 'the charts of the mariners of those seas',[1] and Ramon Lull, writing about the same time, refers to the practice of sailors carrying 'chart and compass',[2] while from Arabic sources we learn that there were already in the twelfth century, if not before, professional writers of nautical instructions: Sahl Ibn Abān, Mohammed Ibn Shādhān and Laith Ibn Kahlān, to name only three.[3] Be this as it may, the complete absence to-day of sailing charts prior to the year 1300 need cause little surprise, for mariners' equipment is liable to be worn out quickly and lost by the men using it. Furthermore, it is unlikely that the multiple production of these charts began much before the fourteenth century.[4] Among the numerous evidences which go to show

[1] *The Book of Ser Marco Polo: op. cit.*, Vol. II, pp. 312–13.

[2] *Vide Arbor Scientiae* (written *c.* 1300), Lyon edition, 1515, fol. cxci. 'Marinarii quomodo mensurant miliaria in mari? . . . Et ad hoc instrumentum habent, chartam, compassum, acum et stellam maris.'

[3] *Vide* G. Sarton: *Introduction to the History of Science*, Vol. II, part 1, p. 221.

[4] *Vide* A. E. Nordenskiold: *Periplus*, pp. 45 *et seq.*

the growing importance of the Portolan chart during this century, not the least significant is the ordinance of the King of Aragon—dated 1354—'that every captain of a ship should carry on board two copies of the marine chart (dos cartas de marear) as well as two rudders and steering wheels'.[1]

Fundamentally, these Portolan charts were the graphical counterpart of the ancient *peripli*, the written itineraries compiled by sailors in classical times from observations made along the coasts they had visited, e.g. the *Periplus* of Scylax of Caryanda, drawn up in the fifth century B.C. Indeed, it is by no means impossible that Italian sailors of the thirteenth and fourteenth centuries used these old manuals as the basis of their maps,[2] bringing them up to date with the help of their own observations based on log and compass readings, that is, on the direct determination of distances and bearings.[3] To distinguish them from the more theoretical works in the Ptolemaic tradition and also from the more imaginative monastic maps, these maps were frequently called compass charts. They contain no indications of longitude or latitude, but they are almost invariably covered with a network of loxodromes, or rhumb-lines, proceeding from a number of foci regularly disposed over the map. While most features of the Portolan chart are slavishly copied from map to map, e.g. the conventional symbols, the use of different colours for different islands and the coastal outlines,[4] the loxodromic net varies in almost every instance. That it was not essential to, or indicative of

[1] *Ordenanzas de las armadas navales de la corona de Aragon, aprobadas por el rey D. Pedro IV, año de* 1354, Appendix No. 1, p. 2.

[2] *Vide infra*, chap. x.

[3] The earliest charts of all may have been constructed without a compass, the azimuths being found from observations of the sun, moon and stars.

[4] A. E. Nordenskiold has ascertained that the places marked in red, for instance, 'are, with only occasional exceptions, invariably the same on all portolanos from the beginning of the fourteenth to the end of the sixteenth century'.—*Periplus:* p. 18. He has also shown that 'as regards the outline of the Mediterranean and the Black Sea, all the portolanos are almost unaltered copies of the same original'.—*Ibid.:* p. 21.

the construction of the map is apparent from the fact that they were normally superimposed upon the chart after the outline and the place-names had been completed—a point that is readily established by an examination of any original. Another difference is that some of the charts have only 16 rhumbs to each focus, e.g. the 'Carte Pisane', while others have 24 and even 32, e.g. the Giovanni da Carignano chart of about the same date, namely *c.* 1300. The most usual arrangement was for a central focus, or rose, to be surrounded by a number of interconnecting and subsidiary roses which would occupy the points of an octagon. All the lines of each rose were produced to the edge of the chart. With the help of these intersections, the pilot could determine, if only very approximately, the relative position of places.

The manner of using the Portolan chart is set out for us in a sixteenth-century book, entitled *The Boke of Ydrography.*[1] Paraphrasing the archaic language, we find that it amounts to this: select a line, in the neighbourhood of the ports of departure and arrival of the day's run, such that it was nearly as possible parallel to the line (imaginary) joining the ports, checking with a pair of dividers: then note which line through the centre of the windrose compass is parallel to the line selected, and so obtain the required bearings.

We see then, that the whole complicated system of lines found on these charts reduces to sixteen superimposed sets of parallel lines. In each set there is one line through a point approximately in the centre of the chart. The sixteen lines through this point form the compass with its 32 'points'; and from it the direction represented by any one of the lines on the chart is determined. In effect, the user of the chart selects that set of parallels which most nearly corresponds in direction with the desired course, and then reads off the designation of the set from the compass.[2]

[1] 'Composed by Jan Rotz, native of Dieppe, in the year 1542.' *Vide* author's own Preface.

[2] N. H. de Vaudrey Heathcote: 'Early Nautical Charts,' *Annals of Science*, Vol. I, 1936, p. 26.

The chart was provided with a circle graduated in quarter-winds (the unit of angular measurement in use by navigators at the time) and as large as the size of the chart permitted. The process of determining the required parallel through the centre compass was made easy by having a sufficient number of parallels ready drawn through these quarter-wind points on the outer circle. The reason for placing the centre of the circle as nearly in the middle of the chart as possible is now self-evident—simply for the purpose of including as large an area as possible within the circle.

The degree of accuracy obtained by such rough-and-ready methods was often surprisingly high. Whereas Ptolemy's figure for the length of the Mediterranean Sea was 20° out, 60° instead of 40°, the Portolan figure is rarely more than one degree out. Outside the Mediterranean, however, the errors grew with a consequent increase in the distortion of the coast-lines. The reason for this was partly the mariner's unfamiliarity with the shores of the open ocean, but partly the use of different scales for the Mediterranean and Atlantic coasts on a number of charts. The original normal Portolan chart was based on a measure of 5,830 metres.[1] The Italian map-makers, however, did not know this unit of length, but they sometimes tried to fit their own mile-measure to the Portolan scale, by assuming that it equalled $\frac{1}{5}$ of a *legua*, i.e. 1,166 metres, whereas the real figure was approximately $\frac{1}{4}$, or 1,480 metres. In actual practice, the latter figure seems to have been employed for maps of the Atlantic, the former for those of the Mediterranean and Black Sea.

The oldest of the Portolan charts to survive are of Italian origin, made at Genoa and Pisa ; those dating from the latter half of the fourteenth century are mainly Catalan. But the typical Catalan map is not, strictly speaking, a Portolan chart. It is more than that ; for while the latter is essentially a sailing guide concerned with coastwise navigation, the Catalan map

[1] *Vide* A. E. Nordenskiold : *op. cit.*, p. 23. This is the length-measure to which Nordenskiold has given the name 'portolan-mile'. It seems to come nearest to the *legua* of the Catalans.

is really a world-map built up around the Portolan chart. It is true that in some cases the term ' world ' connotes simply the habitable earth as known by the author, nevertheless, in others, as the Este Map, it is interpreted to include lands not yet discovered, but only posited. This aggravated the cartographer's task very considerably, for it meant that he was continually being faced with the problem of choosing between scanty and often poorly substantiated fact on the one hand, and plausible and often well-attested theory on the other. It is a tribute to the integrity of these men that their work contains so much that subsequent investigation has proved true. In fact, it is this careful sifting of evidence that constitutes one of the chief merits of the Catalan school of cartography, in an age when intellectual honesty was none too common. The value of the Catalan maps, as commentaries upon the state of contemporary knowledge, at once becomes apparent, and we are hardly surprised to find that the Catalan Atlas of 1375 has the finest delineation of Asia the world had seen up to that time, or that, in its knowledge of Cathay and the Sudan, the same map is surpassed in the Middle Ages only by the Fra Mauro Planisphere of 1459.

Scarcely less valuable, and certainly more interesting for the student of geographical theory, are the Catalan speculations concerning the unexplored territories of the earth. Unlike many medieval scholars, the draughtsmen of Majorca showed a praiseworthy restraint in this respect. Thus we may look almost in vain for those fanciful creatures with which the cosmographers of that age filled their empty continents. At the same time, these men saw nothing strange about a belief in the Terrestrial Paradise or in a hydrographical system stretching from sea to sea.[1] For the most part their speculations were of another kind, and usually they contained at least a part-truth. For instance, ' Lacus Nili '—the ' Pactolus ' of Strabo and the ' Palolus ' of later maps—which, in the Catalan Atlas and subsequent works is located in the neighbourhood of Timbuktu, may reasonably be identified with the flood region of

[1] *Vide* Plate No. IX.

the Niger River above that town. Again, in the anonymous map of *c.* 1440, the presence, south of 'Palolus', of the legend 'Montanie del Lor', where the five feeders of that lake have their head-waters, is a possible attempt to show the Fouta Djallon mountains of French Guinea. In this connexion it is not without significance that this range of mountains has five main tributaries, namely, the Upper Gambia, Faleme, Bafing, Tinkisso and Niger.

On one matter the map-maker could hardly refrain from speculating, for this reason: land exploration had for a long time now outrun oceanic discovery, and so, concerning Africa, for example, much more was known of the Sudan by the end of the fourteenth century—as we have shown in an earlier chapter[1]—than was known of the oceanic fringe in the same latitudes. The earlier draughtsmen insisted upon cutting the continent short just beyond the limit of coastal knowledge, that is, in the vicinity of Cape Bojador. By so doing, however, they found themselves reducing the vast extent of the Sahara almost to vanishing point. Thus, in the map of 1375, Sigilmesa and the Rio del Oro (i.e. Senegal-Niger system) are placed in closer proximity than Ceuta and Cape Non. Later draughtsmen, in order to escape the embarrassment caused by indicating the great trans-Saharan caravan routes within these narrow limits, began to speculate on the course of the west African coast, south of Bojador. By general agreement it was made to trend south-south-east.[2] Speculation of this sort did at least have the merit of enabling the map-maker to draw the Sahara with greater accuracy. In passing, it should be noted that all the Catalan maps, with the exception of the Este World Map, which was the last of its line, stop short on their southern side in the latitude of Sierra Leone approximately, that is, where knowledge gave place to ignorance.

Associated with this habit of speculation, was the practice of 'harmonizing' established facts with traditions—a practice which became very popular from the fourteenth century onwards. The people who found pleasure in reconciling the

[1] *Vide ante*, chap. v. [2] *Vide* Plate No. VII.

views of Pliny and Ptolemy, of Aristotle and Ambrose, were not easily disturbed by the challenge of the new school of practical cartography. On the contrary, with real but heedless enthusiasm, they set about the task of pouring the new wine into the old skins—an occupation offering more and more difficulties as exploration extended the known world. Not even the Catalan cartographers could avoid it. But as they were predisposed to eschew wild guesses and to be sceptical of travellers' tales, their maps do not afford the best illustrations of this characteristic. We will confine ourselves to a single example. At the beginning of the Catalan period the 'Rio del Oro'—a heritage of classical geography—was made to debouch into the Atlantic immediately south of Cape Bojador. With the extension of trans-Saharan commerce in the fourteenth century, and, along with it, the enlargement of geographical knowledge, the 'Rio del Oro' was pushed, little by little, farther south, until at length, in the Este Map, it is located approximately in the latitude of the Senegal-Niger system, which, no doubt, it is intended to represent.

Prior to the mid-fifteenth century, this 'harmonizing' problem presented few serious obstacles. Maritime exploration had hardly begun to yield fruit, while the land explorations of the Polos and their contemporaries had not yet produced a systematic revision of current ideas. Where the results were found not to square with the ideas, as was, of course, usually the case, it was quite customary to find them being either distorted beyond recognition, or neglected altogether. Thus, in his geography of Abyssinia, Fra Mauro, the great Venetian cosmographer of the fifteenth century, weaves Polo's narrative into Arab theory, and makes these together fit the topographical notions of Abyssinia which he had obtained from first-hand sources. Again, while professing knowledge of Polo's Asiatic wanderings, Mauro does not even hint at the peninsular nature of South India, the fact of which is implicit in Polo's statements.

With the development of Portuguese seafaring in the fifteenth century and the consequent widening of the southern

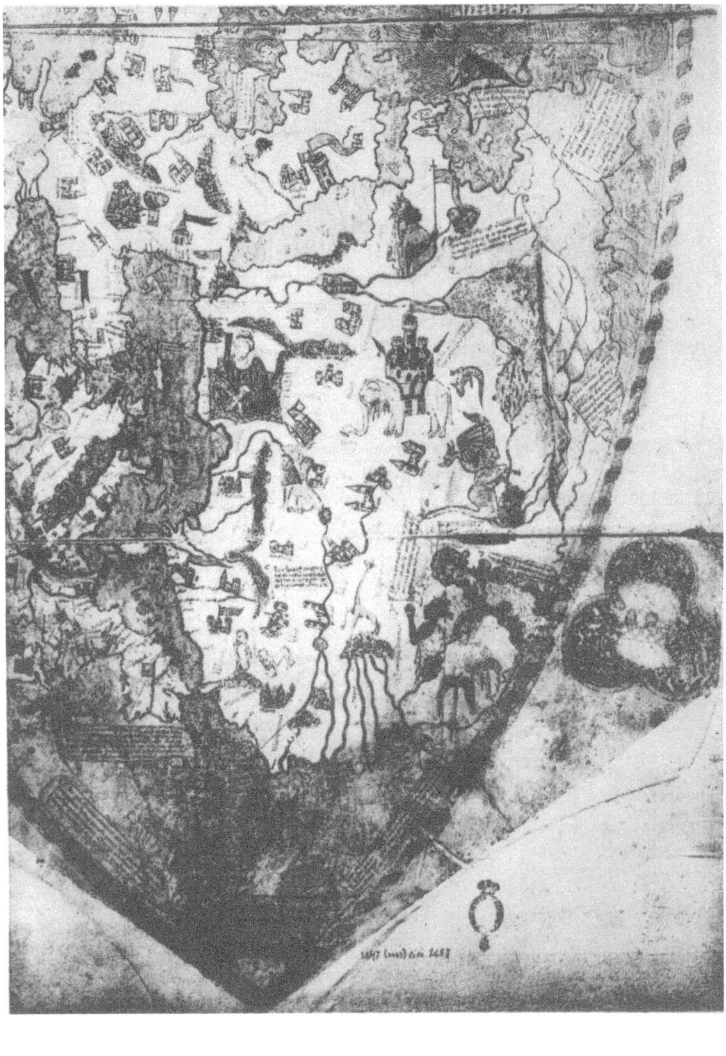

XVII

AFRICA—ACCORDING TO THE GENOESE WORLD MAP (c. 1457)

(Biblioteca Nazionale, Florence)

horizon, the 'harmonizing' problem became increasingly acute. Each map-maker tackled it *de novo*, so that scarcely any two world-maps of this period give us the same world-view. Compare, for instance, the Este, the Walsperger and the Genoese maps,[1] all of approximately the same date, *c.* 1450. In the first, at least three distinct influences, in addition to that of the Portolan chart, may be detected. They are Classical, Christian and Arab. Of these only the Arab is strong, while it is improbable that the Classical influence is direct. Thus, the map owes nothing to Ptolemy, and it is less likely that its author should have taken his idea of a southern continent direct from Crates, the originator of the concept, than that he should have taken it from Arab or Christian cosmographers, such as Abū'l Fidā or Isidore, who revived it. The influence of the medieval Christian tradition is betrayed in such things as the legend relating to Prester John and the portrayal of the Terrestrial Paradise. There can be no mistaking the Arabic influence. We have only to compare the delineation of the southern half of Africa on the map with the description given by the eleventh-century writer, Al-Bīrūnī, of the shores of the Southern Ocean to be convinced of the kinship.[2]

In the Genoese Planisphere, the Portolan chart again serves as the pattern for the Mediterranean region, but elsewhere it is the Ptolemaic tradition which is mostly invoked. Practically all the features of the Sahara are similar in their territorial relationships and outlines to those occurring on a Ptolemy manuscript of 1400 in the Laurentian Library at Florence. The same influence, though not the same slavish adherence, is apparent to the south of the Sahara, where the author places a great gulf, containing an island and a legend, now barely visible. When translated, it reads: 'Contrary to the opinion of Ptolemy, this is a gulf, but Pomponius [Mela] speaks of it with its islands.' Or again, in another rubric:

> Beyond the equinoctial line Ptolemy records an unknown land, but Pomponius [Mela] and many others as well raise a doubt

[1] *Vide* Plates Nos. IX, XVI, XVII. [2] *Vide ante*, chap. iii.

whether a voyage is possible from this place to India ; [neverthe-less] they say that many have passed through these parts from India to Spain . . . especially Pomponius in his last chapter.

This rather refreshing disposition towards agnosticism is exemplified again in the configuration of South Africa. Here the author does not follow the Alexandrine scholar in attributing to it an eastern prolongation, but contents himself with rounding it off in the conventional way, namely, in the form of a half-moon. The division of geographical loyalties at this period is further illustrated in the placing, side by side, of the Nile sources—located in the 'Montes Lunae', after Ptolemy —and this typically medieval legend : 'Some have represented the Paradise of Delights in this region, while others have said that it is beyond the Indies to the East . . .'

In the Walsperger Map [1] the author, a German, will have nothing to do with modern practice as expressed in the Portolan chart, utilizing in its stead the recently exhumed *Geography* of Ptolemy. Into this mould he pours the usual ingredients for a medieval *mappamundi*. This attempted fusion of classical and ecclesiastical ideas produces results at once artless and unsatisfactory. Consider Africa, for instance. The western littoral starts off with a plainly Ptolemaic trend past 'Hesperidum' as far as 'primum clyma Meroys'—one of Ptolemy's seven *climata*. Here it turns east, past the country of the *Egibani*, who boast the form of goats, and that of the *Sciapodae*, conspicuous for the size of their feet. These are Plinian in parentage. At this point the coast turns southwards again to the edge of the map, near which we read the most un-Ptolemaic observation that 'Around this pole there are most wonderful creatures, not only beasts, but men.' The eastern prolongation of the continent, extending as far as 'Java Insula', and separated from Asia only by a narrow strait, once again brings us back to the true Ptolemaic tradition, as also the placing of the source of the Nile in the heart of Africa.

As for the rest of the world, it is thoroughly medieval in

[1] *Vide* Plate No. XVI.

sentiment. Fancy runs riot and facts are badly distorted. Two Nilian lakes, 'Lacus Meroys' and 'Lacus Affrorum', are given the dimensions of Iberia. The rivers, four in number, flowing northwards from the Atlas Mountains are each longer than the Elbe and Oder. The stock-in-trade of the theologian is capitalized to furnish the author with a Terrestrial Paradise and its usual perquisites. Jerusalem, in conformity with the popular belief, is placed in the centre of the earth.

Notwithstanding their essential individuality several maps of this period, especially those dating from the second half of the fifteenth century, bear witness to the operation of a new factor. This is proved by the following facts. The first is the scarcity of maps portraying the Portuguese discoveries. That the Portuguese had maps and made them cannot be doubted when we recall the numerous allusions to them in contemporary literature. Alfonso's statement, for example, that, until Prince Henry's time 'no one knew anything of the land beyond Cape Bojador, nor was it marked on the sailing charts and *mappaemundi* save as men pleased,'[1] and Fra Mauro's, to the effect that the Portuguese had 'framed new charts' of the west African coasts, both tend to confirm this belief. There are, moreover, frequent references to Portuguese maps in Zurara's *Chronicle of the Conquest of Guinea*[2] and Pacheco's *Esmeraldo de Situ Orbis.*[3] How, then, are we to account for the fact that only the merest handful of fifteenth-century maps, showing the results of the Portuguese discoveries, survives, and also for the entire absence of Portuguese maps in that century and for their rarity in the early decades of the next?

The second fact is the indifferent quality of two of the most famous examples of fifteenth-century cartography, Fra Mauro's Planisphere and Martin Behaim's Globe.

Now, we learn from an authoritative source that for making the Planisphere Mauro was supplied by the Portuguese

[1] *Alguns documentos do archivo nacional da Torre do Tombo acerca das navegaçoes e conquistas Portuguezas*, p. 8.

[2] Chap. 78.

[3] e.g. I, 5, and I, 13 (Hakluyt Society).

with 'information of all new discoveries made or projected',[1] in return for his professional services in drawing it. By 1459, the year of its construction, the Portuguese had sailed some 2,000 miles beyond the Straits of Gibraltar, that is, as far as Rio Grande.[2] Mauro knew this, for he tells us as much in a rubric near the west coast of Africa and adds circumstantially that 'everywhere they found the coast not dangerous, with the soundings good, convenient for navigation and with no risk from storms'. He continues: 'they have framed new charts of these regions and given names to the rivers, bays, capes and ports. I have many of these charts in my possession . . .' Of such extensive knowledge, Mauro's west African coast offers barely a trace.[3] Beyond Cape Roxo there is no linear correspondence with the actual coast-line and the only Portolan chart bearing comparison with Mauro's map at all is one by Andrea Bianco—a Venetian, known to have been one of Mauro's collaborators. Now assuming that he was telling the truth (Mauro had nothing to gain by lying and, moreover, he shows himself elsewhere in his work to have been a faithful follower of his authorities), we are reluctantly forced to the conclusion that the charts given to him were worthless counterfeits and that the latest findings of the Portuguese had been suppressed.[4]

The study of Behaim's Globe points to a somewhat similar conclusion.[5] On investigation, it will be found:—(*i*) That the only map extant which resembles the globe at all closely in its west African nomenclature is the Martellus Germanus World Map of *c.* 1489. In fact, 80 per cent of the Martellus place-names, as far south as Monte Negro—beyond which point no correlation is possible—figure almost unchanged on the globe. (*ii*) That it is impossible to detect direct Portuguese

[1] H. E. Wauwermans: *Histoire de l'école cartographique Belge et Anversoise du XVIe siècle*, Vol. I, p. 135.

[2] i.e. Jeba River, 12° N.

[3] *Vide* Plate No. XII.

[4] *Vide* G. H. T. Kimble: 'Portuguese Policy and its Influence on 15th-century Cartography,' *American Geographical Review*, Oct. 1933.

[5] *Vide* Plate No. XVIII.

XVIII
AFRICA—ADAPTED FROM MARTIN BEHAIM'S GLOBE (1492)
(Nürnberg)

influence, either on Behaim's place-names or on his delineation of the coast-line, for only 40 per cent of the 120 legends appearing along the western littoral occur, even in a modified form, in the roughly contemporary *Guinea Portugalexe* and La Cosa maps, which may be taken to be Portuguese in parentage, if not in actual execution. (*iii*) That at least one half of the west African names have no known source. If these names are *bonâ fide*, it is almost certain that they must have been derived from a Lusitanian source, especially as Behaim was closely associated with the life of Lisbon from 1484 to 1490. If this were the case, however, it is difficult to see why the names remained entirely unutilized by later map-makers—a difficulty all the greater if, as Behaim averred, he accompanied Diogo Cão on his second voyage of 1484–5 when the information would have been secured. Have we, then, to assume that Behaim invented 50 per cent of his nomenclature? This conclusion, although unsavoury, appears unescapable, but we can readily understand such a fabrication if the Portuguese kept back the information he was needing. In Germany people were asking for a picture of the world that would incorporate the results of the recent discoveries. Was Behaim to confess that the Portuguese had denied him the details, or was he to invent them? The desire to masquerade before his fellow-countrymen as a successful explorer as well as a great traveller appears to have dictated the answer.[1]

The third fact pointing to the operation of a new influence is the time-lag between discovery and mapping.

The earliest unequivocal map record of Portuguese discovery is provided by the Andrea Bianco Portolan chart, dated 1448, that is, fifteen years after the rounding of Cape Bojador. (It is worthy of mention, however, that this map gives the results of very recent explorations, including the discovery of Cape Verde.) The next nineteen years proffered

[1] For a fuller discussion, *vide* G. H. T. Kimble: 'Some Notes on Mediaeval Cartography', *Scottish Geographical Mag.*, Vol. XLIX, 1933, pp. 95 *et seq.*

nothing, and when in 1467 Gratiosus Benincasa produced the first of his famous series of charts, it resembled the Bianco map so closely that it must have been copied either directly from it or from a common original. It suggests, moreover, that very little knowledge had filtered through to Italy for nearly twenty years; yet, during these years, Cadamosto and his companions had pushed down the coast to within 6 degrees of the Equator. It was not until 1468, fourteen years after the actual discovery, that a chart, also by Benincasa, incorporating the findings of the Cadamosto expeditions to the Senegambia territory, was forthcoming.[1]

How are these three facts to be accounted for? It has been shown by certain Portuguese scholars [2] that, in pursuance of their ambition to hold a monopoly of the trade of West Africa, successive kings of Portugal decided on the suppression of all information calculated to excite the interest and jealousy of other powers. Further, that vessels of other nations were prohibited from sailing into West African waters. Duarte Pacheco tells us of one expedition made by some Flemings in 1475 to Mina in defiance of this veto and adds that their folly was suitably rewarded, for 'God gave them a bad end'.[3]

John II (reigned 1481–95) was the instigator of this policy, using his energies to prevent leakage of the news of discoveries at a time when foreigners were seeking by every means to acquire it. In the reign of his successor, Manuel I, the vigilance of the Government was intensified, especially after the return of Cabral from India. 'It is impossible to get a chart of the voyage,' wrote an Italian agent concerning Cabral's expedition, 'because the King has decreed the death penalty for anyone

[1] With the speeding up of maritime activity along the African coast in the later decades of the century, the 'time-lag' ceased to be noticeable as far as the incorporation of the new discoveries in maps is concerned. Thus the result of Diogo Cão's first voyage of 1484 was made public before the end of the same decade in the *Guinea Portugalexe*—a Venetian compilation.

[2] e.g. J. Cortesão: *Vide Lusitania*, Jan. 1924.

[3] *op. cit.*: II, 3.

sending one abroad.'[1] It is also said that charts were sometimes only lent to navigators by the Portuguese India House and at the end of a voyage had to be returned to that institution.[2] Again, evidence is not wanting that official documents were sometimes tampered with and had their maps confiscated. Pacheco's *Esmeraldo* . . . , containing approximately twenty strip maps of the coast, besides numerous sketches and a world-map, disappeared from contemporary literature and survives to-day only as a half-complete work devoid of illustrations. Rather significantly, the manuscript comes to an end just where the author is about to give instructions for the navigation of the sea route to Calicut from the Cape of Good Hope. Just as significant is the fact that the first subsequent notice of the work dates from the middle of the eighteenth century, that is, long after the danger of its being capitalized by foreign interlopers was past. Accordingly, there seems to be little or no doubt that the work was confiscated, for it is unthinkable that a document—the only one of its kind then in existence describing in detail the navigation of 10,000 miles of ocean coast—should escape the eagle eyes of the Portuguese invigilators.

The answer to our question, therefore, would appear to lie in the realm of Portuguese colonial policy—in the 'conspiracy of silence', as it has been called—of the late fifteenth and early sixteenth centuries. But such a view is not entirely free from controversy, for it will immediately be asked, for instance, what could have been the object of the Portuguese in withholding from Mauro information which would have enhanced the value of his Planisphere, and in substituting for it presumably worthless caricatures of charts? In attempting to solve this riddle we need to realize that the chief cartographical need of the Portuguese at the time was not so much a map of the world that would merely portray with precision what they already knew, as a map that would give the opinion of

[1] *Vide* L. P. da Silva: *Historia da colonisaçao Portuguesa do Brasil*, Vol. II, p. 227.

[2] *Vide* E. Prestage: *Portuguese Pioneers*, p. 171.

geographers on the extent of Africa and on the possibility of a sea-route to the East. Mauro's Planisphere fully met this need. At the same time, by its allusions to navigators in the southern seas,[1] it was calculated to spur on the Portuguese to renewed effort to reach their goal.

Whatever the true explanation of these and other features of the planisphere may be, the importance of the work, for us, is largely unaffected. In an admirable way it illustrates, as also does Behaim's Globe, the division of geographical loyalties and the state of world-knowledge on the eve of the Great Age of Discovery.

[1] *Vide ante*, chap. v, p. 118.

CHAPTER IX

THE GEOGRAPHICAL BACKGROUND OF THE GREAT AGE OF DISCOVERY

AFTER the time of Albert the Great and Roger Bacon the pace of scientific progress in the West slackened temporarily. The Europe of the Black Death and the Hundred Years' War still provided asylum for the devotee of settled life and quiet study, but it did not recapture the atmosphere of the thirteenth century which had seen scholasticism reach its height, nor did it produce as many notable men of learning as that century. All the same, the age was not one of decline or even of stagnation, for there was much intellectual activity. Aristotelian physics were being attacked by William of Occam [1]; astrology and magic by Nicolas Oresme [2]; the reform of the calendar was being advocated by Clement VI and Pierre d'Ailly, and the 'Newtonian' laws of gravity were being adumbrated in the writings of Albert of Saxony.[3] Simultaneously, changes were taking place in the philosophical outlook. Most significant of these was the loosening of the scholastic grip upon secular studies. The men to whom the credit for this belongs are, pre-eminently, Duns Scotus and Occam, for these men first openly repudiated the doctrine that all knowledge could be comprised within the one Truth as it was understood by the Church, accepting, in its stead, the principle of the twofold nature of truth, namely, the acceptance by faith of the Christian dogmas and the examination by reason and

[1] *Vide* his Commentary on the *Sentences* of Aristotle.

[2] *Vide, inter alia, Des divinations, c.* 1361, and *De Configuratione Qualitatum, c.* 1370.

[3] *Vide Subtilissimae quaestiones in libros de Caelo et Mundo.*

experiment of philosophical and scientific subjects. Naturally, the new doctrine was violently opposed by the Church; Occam himself was imprisoned for a time. Nevertheless the doctrine spread, but it was not until the end of the fifteenth century that the last show of ecclesiastical resistance was abandoned. Thenceforward men of learning were better able to advance their inquiries free from the obligation to reach conclusions fore-ordained by theology, and theology, being at long last detached from science, was given an opportunity to develop its no less important mystical aspects.[1]

The exponents of scholasticism occupied, as we have seen, the rôle of interpreters; original experimental investigation was, for the most part, foreign to their ideas. At the same time these men, nimble intellectuals and masters of dialectic that they were, kept alive and even intensified the spirit of logical analysis, while, as Dampier-Whetham has reminded us, 'their assumption that God and the world are understandable by man, implanted in the best minds of Western Europe the invaluable, if unconscious, belief in the regularity and uniformity of nature',[2] without which scientific research would never have been attempted. As soon as they had begun to throw off the shackles of ecclesiastical authority, these men of the Renaissance set about using the lessons which scholastic method had taught them. They began observing

in the faith that nature was consistent and intelligible, and when they had framed hypotheses by induction to explain their observations, they deduced by logical reasoning consequences which could be tested by experiment. Scholasticism had trained them to destroy itself.[3]

This new movement—Humanism, as it came to be called—was born in Italy of a new feeling for the greatness of ancient Greece and Rome. There, as manifested in the works of Petrarch and his followers, one of its first and greatest mani-

[1] Unfortunately, this is precisely what it did not do. It began, instead, to exhume the corpse of thirteenth century scholasticism.

[2] *A History of Science*: p. 104. [3] *Ibid.*

festations was, paradoxically, not a love of novelty, whether in literature, science or art, but a return to a love of the real classics. It was now seen that there had been a time when men had used all their faculties of thought and expression without fear or favour. Unrestricted by dogmas and untrammelled by religious scruples, these men had bequeathed to posterity a literature different in quality and range from anything that had been composed during the intervening centuries. This literature was now welcomed not only as supplying new and higher literary standards, but as disclosing a new conception of life, one which gave large scope to the play of the emotions, to the sense of beauty and to all the activities of the intellect.

The acquisition of this new learning was effected by a small body of accomplished Greeks who were encouraged by rich patrons to settle as professors in various Italian towns, and by the visits which certain Italian scholars made to Constantinople for the purpose of studying the Greek language there. The first real teacher of Greek in Italy—' the man with whom the revival of Greek learning in the West began '—was Manuel Chrysoloras, who lectured on Greek at Florence, Pavia, Milan and Venice, *c.* A.D. 1400. The movement so vigorously inaugurated by him was continued by several of his compatriots, most of whom came to Italy between 1400 and the fall of Constantinople in 1453.[1] The conquest of that city by the Turks did, of course, serve to hasten the spread of classical culture, for it led to the westward dispersion of many competent teachers who brought manuscripts with them to their new homes. The search for manuscripts quickly became a fashionable pursuit among the wealthy, not only in the Levant but also in the monastic and cathedral libraries of Italy and Western Europe. The works thus acquired formed the basis of many a famous library whose origin dates back to the fifteenth century. Nicholas V founded the Vatican library in 1453, Cosmo de Medici began the

[1] The restoration of Greek letters in Italy preceded the downfall of the Eastern Empire, and was not, as has sometimes been supposed, a result of emigrations caused by that event.

Medicean collection a little earlier, while his contemporary Poggio Bracciolini ransacked all the cities and convents of Europe in search of ancient writings. Giovanni Aurispa, a pupil of Manuel Chrysoloras, brought home 238 Greek codices from Constantinople on a single journey. By such men manuscripts were worshipped just as the relics of the Holy Land had been worshipped by their great-grandfathers.

So long as the only method of re-duplicating these books consisted in the laborious process of copying them on an inconvenient and expensive material, parchment, the possession of a library was only within reach of a favoured few. It was not until the invention of movable type, in the middle of the fifteenth century, had transformed the earlier attempts (in the same century) at printing with fixed moulds into a practical proposition, that books became more abundant.

Simultaneously, a renewed ardour for geographical discovery manifested itself, and, in spite of the primitive state of the art of navigation,[1] the area of *terra cognita* was rapidly extended.[2]

How did these new expressions of an intellectual renaissance affect geographical thought? Again, paradoxically enough, more by a deepening of interest in the classical than in the contemporary world. In the huge encyclopaedia, *Fons Memorabilium Universi*, of Dominicus Bandinus of Arezzo [3]—a work by no means devoted exclusively to geography or geographers—men like Archimedes, Aristotle and Agathocles are given more than a page apiece, but Marco Polo is dismissed in three lines, and as for Abelard and Albertus, they are not so much as mentioned. Writers of more strictly geographical works, such as Pierre d'Ailly, Cardinal of Cambrai in the early fifteenth century, reveal a similar preoccupation. Although d'Ailly compiled his *Tractatus de Imagine Mundi*

[1] *Vide infra*, chap. x.

[2] For the history of maritime enterprise in the fifteenth century, *vide, inter alia*, E. Prestage: *Portuguese Pioneers* and A. P. Newton (editor): *The Great Age of Discovery*.

[3] *Fl. c.* 1400 A.D.

140 years after the time of Roger Bacon (who mentions William of Rubruck, as we have seen) he does not name a single medieval explorer, whether Asiatic or African. That all knowledge of recent explorations should have escaped the Cardinal's notice is hard to believe. Had not his own countrymen, Jean de Bethencourt and Gadifer de la Salle, for instance, attempted to conquer and evangelize the Canary Islands at the very beginning of the fifteenth century? The whole work, with the exception of a few Arabic citations, mainly from Latin translations of the works of Avicenna and Averroes, might have been composed a thousand years previously. Indeed, some of it had. To take only one instance—his discussion of the Antipodes.[1] In approaching this controversy, d'Ailly states it with such clarity that we expect to hear him assert the truth. But as a good theologian he finds himself confronted by the arguments of St. Augustine and the biblical texts upon which it was founded.[2] Overpowered by the weight of this authority, d'Ailly takes refuge in hearsay.

> Some say that it (i.e. the zone between the winter Tropic—Capricorn—and the Antarctic circle) is as temperate and habitable as our own . . . However, according to them, there can be no communication between the Antipodes and us because of the impossibility of crossing the intervening Torrid Zone and the Tropics. According to this opinion, the population of this region would be ignorant of the teaching of Christ and the Apostles, which is contrary to the sacred affirmation that 'their sound went into all the earth and their words unto the ends of the world'.[3] In the sixteenth chapter of his *City of God*, St. Augustine refutes this opinion. However, certain writers maintain that it is a fable and that the fourth zone (i.e. the southern temperate zone) is for the most part covered with water, and this view is supported by highly probable reasons . . .

The reasons given, however, such as the claim that the elevation of the sun there makes life impossible—presumably because it scorches the earth—are highly improbable. For d'Ailly's

[1] *Tractatus de Imagine Mundi*: cap. 7.

[2] *Vide ante*, chap. ii.

[3] *Romans* x. 18.

personal opinion on this subject, or for that matter on any other, we shall search in vain.

That d'Ailly was aware of the shortcomings of his *Imago Mundi* may, perhaps, be inferred from the appearance of a subsequent treatise. This was the *Cosmographiae Tractatus Duo*, written about 1414, by which date a Latin copy of Ptolemy's *Geography* had come to his notice.[1] In the earlier treatise, *c.* 1410, d'Ailly knew of Ptolemy neither as a geographer nor cosmographer, but only as the author of the astronomical *Almagest* and the astrological *Quadripartitum*. Finding himself in possession of a veritable mine of new, if in another sense ancient, material, he is not slow to exploit it. At the same time he makes no attempt to incorporate all the Ptolemaic technique into his own system. Thus he finds the twenty-two parallels of the Alexandrine climatic division too cumbersome for his own use and reduces them to twelve. Seven of these correspond to seven of Ptolemy's *climata*; three, namely, Montes Barditi, Raptum and Cattigara, are adopted from the *ante-climata*, while the remaining two, Britain and Thule, pass through the *post-climata*. Out of the thousands of places tabulated in the *Geography*, d'Ailly selects 353, the choice of which hardly commends itself from a geographical standpoint, giving them their Ptolemaic latitude and longitude. Whereas Meröe was formerly d'Ailly's most southerly outpost of the *oikoumenē*, now it is *Montes Barditi qui meridionalem limitem nostre habitabilis terminant*, and to these he gives the latitude of 16° S. Ethiopia is similarly affected, receiving an ill-defined southern extension and made to

[1] The translation of this work into Latin was primarily due to the famous Italian patron of letters, Palla Strozzi, who brought over Chrysoloras to Florence. Chrysoloras himself seems to have begun the translation, but not to have completed it. This was done by his pupil Jacobus Angelus in 1406. However, Angelus dealt only with the text: the maps were taken in hand by two almost contemporary Florentines who multiplied copies of the text and the maps in the same large format as the Greek originals. *Vide* Joseph Fischer: *Claudi Ptolemaei Geographiae Codex Urbinas Graecus* 82 . . ., Vol. II *passim*, and E. L. Stevenson: *Geography of Claudius Ptolemy*, Introduction (by J. Fischer), pp. 3–15.

adjoin the hitherto unmentioned land of Agysimba. Again, whereas in his first treatise d'Ailly describes Africa *parem Europae longitudine, tamen multo angustiorem*,[1] now he tacitly refutes it by admitting it to be habitable as far as 16° S. and to extend still farther.

But it must not be imagined that the Cardinal followed his new-found authority unquestioned. Occasionally he leans towards the arguments of Pliny and others, and this, so it would seem, less from caprice than from an appreciation of their intrinsic merit. Thus, for instance, in a passage made famous by its association with the name of Columbus, he demonstrates that the length of the habitable earth is greater than that postulated by Ptolemy and points out, as a corollary, that the commencement of India in the East could not be very far distant from the western extremity of Africa.[2] From the same passage we learn that d'Ailly favours the concept of an open, rather than an enclosed Indian Ocean, and, in consequence, a sea-girt Africa, although this is nowhere stated categorically.

When all is said and done, however, d'Ailly is nothing if not deferential. There is not a single idea in any of his geographical writings which cannot be traced back, sooner or later, to some impeccably orthodox source. At the same time there is a keen sense of proportion pervading them all and an obvious sincerity of intention. Thus, the Cardinal steadily refuses to have anything to do with such superstitions as the monstrous one-eyed giants—the *Monoculi*—and the

[1] *Tractatus* . . ., cap. 32.

[2] Cap. 19. . . . 'Multo major est longitudo terrae versus orientem quam ponat Ptolomeus, et secundum philosophos Oceanus qui extenditur inter finem Hyspaniae ulterioris, id est Africae, a parte occidentis, ad inter principium Indiae a parte orientis, non est magnae latitudinis. Nam expertum est quod hoc mare navigabile est paucissimis diebus si ventus sit conveniens . . . Secundum vero quod protendunt autores et maxime Plinius, mare indicum decurrens per latus meridianum . . . indie. Deinde immensum terre spacium absorbens se flectit fauces maris rubri et mare ethiopum ubi incipit Ethiopia.' Cf. R. Bacon: *Opus Majus*, Part IV, p. 329 (R. B. Burke's edition.).

'umbrella-footed' folk—the *Sciapodae*—who were commonly reputed to be inhabitants of central Africa. On the other hand, it is regrettable that his preoccupation with classical erudition should have blinded him to the importance of events nearer his own time.

Unfortunately, this preoccupation with the past was not confined to d'Ailly. New anthologies of 'historical' geography were constantly forthcoming and, after the introduction of the printing press, enjoyed a quite phenomenal circulation. Caxton himself printed at least two such works, *The Mirrour of the World*[1] and the *Polychronicon*.[2] Look where we may in either, it is impossible to find an original statement. If the fact that both these books ran into several *incunabula* editions is any criterion of the diffusion and acceptability of their contents, then it is plain that contemporary events had as yet done little to modify the long-standing geographical background.

> Affryca conteyneth many provynces and landes. Fyrst it conteneth the West dele of Ethiopia, thenne libia, tripolis, getula, numidia and two Maritania . . . Ethiopia hath thre partyes, the fyrst is hylly and montuous and stretcheth from the mount Athlas unto Egypte. The myddel partye is ful of gravel. The thyrdde, that is, the lest partye is almost all wyldernes, that partye is bytwene the South Ocean and the Ryver Nylus and hath the Reed see on the east syde. . . . In Ethiopia been many dyverse peple wonderly and grysly and shapen . . . Somme curse the sonne for his grete heete. Somme ete serpentes and addres, somme hunt lyons and panters. Somme dygge caves and dennes and dwelle under erthe . . .[3]

Although this description was written a century before it was printed in the year 1480, Caxton passed it for current knowledge.

The scholarly Pope Pius II (Aeneas Silvius) writing about the time Caxton was printing, shares the prevailing ignorance

[1] Translation of the anon. *Image du Monde*, *vide ante*, chap. iii.

[2] Composed in the fourteenth century by a monk of Chester, Ranulf Higden by name.

[3] *Polychronicon*: cap. 19.

of contemporary exploration. Africa, it is true, is not his concern.[1] What little he does say about it in his introduction is, however, not without value, if only for its indictment of Ptolemy's theory of the Indian Ocean. In this he exhibits a welcome independence of thought. On the other hand, there is still a lamentable ignorance of (or it may be merely lack of interest in) current affairs.

> By common consent [he assures us] it is held that in the whole of the habitable world there exist three principal regions, of which, in point of size, Asia is the first, Africa the second and Europe the third. Asia joins Africa (as was shown by Ptolemy) by the back of Arabia which separates our sea (i.e. the Mediterranean) from the Arabian Gulf. No one denies this, but he (Ptolemy) adds that in a certain place, it is joined by unknown land which encompasses the Indian Ocean. In this opinion he is almost alone. For all whom we meet writing about the features of the earth place the Indian Ocean to the South and East, without ascribing to it any bounds, being of the opinion that it forms part of the (open) Ocean—as has been handed down by those who have navigated from the Arabian Gulf to the Atlantic Sea and the Columns of Hercules.[2]

Elsewhere, as in the chapter[3] on the circumnavigation of Africa, or where he deals with the question *An sub aequinoctiali circulo ac torrida zona sint habitationes*,[4] Pope Pius employs the same body of evidence that has served his predecessors, and comes to almost the same conclusions. Ptolemy, as we should expect from the date of the work, is given very considerable prominence. His arguments, all the same, seem to leave the Pope for the most part unconvinced. On the question of the circumnavigability of the continent he differs from his Alexandrine master on *à priori* grounds, while on that of the habitability of the Torrid Zone he is content merely to re-state Ptolemy's view, leaving all criticism of it to the reader.

[1] The work is styled *Asiae Europaeque elegantissima descriptio* . . . and was probably written during his tenure of the Papal office, i.e. 1458–64.

[2] *Ibid.:* cap. 6. This is a reference, no doubt, to the expedition promoted by the Pharaoh Necho, *c.* 600 B.C.

[3] *Ibid.:* cap. 2. [4] *Ibid.:* cap. 5.

Pope Pius was wrong, however, in supposing that there was little or no contemporary support for the Ptolemaic notion of an enclosed Indian Ocean. As we have already seen [1] more than one cartographer of the fifteenth century adopted a view which approximated to it. In the text-books of the time, we find, among others, Giovanni da Fontana re-fashioning it in a form which was really a compromise between the Arabic and the Alexandrine. He held that that Ocean was practically landlocked, being bounded on the east and south by an unknown landmass which was partly accessible and through which, so he opined, there was a passage connecting the waters of the Indian Ocean with the outer sea.[2] The former, in his view, formed part of a vast midland sea which embraced the Mediterranean and the other 'enclosed' seas. The latter was simply part of the all-encircling Ocean 'stream'.

It is evident from the foregoing facts, therefore, that Ptolemy was enjoying a considerable measure of popularity in the later decades of the fifteenth century. Not that he was a new name to the medieval student of geography. Ever since the twelfth century (Gerard of Cremona's translation of the *Almagest*—the first—dates from 1175) Latin versions of his mathematical and astrological works had been appearing with remarkable rapidity. While all these works contributed something to the esteem—reverence is the better word—in which he came to be held, the *Almagest* exerted by far the greatest influence on medieval scientific thought. Summing up, as it did, the labours of Ptolemy and his predecessors in systematic and comprehensive form, it rapidly acquired, for Saracen and Christian alike, the character of a higher revelation. Indeed, the name by which it is known is commonly explained

[1] *Vide ante*, chap. viii.

[2] *Vide Liber de Omnibus Rebus Naturalibus* (1554 Edition): III, 18. '. . . mare Indicum quod est in latere orientali australi interclusum ex parte sua septentrionali terra Indica et occidentali terra Ethiopica, ex oriente vero et austro terra incognita est pro parte accessibili. De quo nostri ferunt manifestium hostium non habere, sed nos credimus illud habere et aditum esse ad oceanum vel brachium Amphitritis meridionale sed fortasse hominibus ignotum.'

as a superlative title.[1] And what this work was for astronomy, Ptolemy's *Quadripartitum* (or *Tetrabiblos*) was for astrology. Early translated into Arabic, it was widely read among the Moslems and became the subject of several commentaries. Naturally it was one of the earliest works to be turned into Latin—the version of Plato of Tivoli is dated 1138—for astrological prediction stood high in public favour, producing some of the 'best-sellers' of the *incunabula* period and absorbing no small fraction of the time of medical faculties. The effect of the re-discovery of Ptolemy's *Geography*, authentic, convincing and vastly superior to the average text-book of the day, can therefore be imagined. A genuine Shakespearian find would cause no greater stir to-day. But powerful though it was, Ptolemy's influence upon geographical thought never succeeded, as that of Aristotle and others had previously done, in paralysing independent inquiry. The recovery of his *Geography* had been effected too late (1406) in the history of man's intellectual emancipation to bring his mind back into a new servitude. Coming when it did, its effect was stimulating rather than enslaving. This is seen very clearly in connexion with 'Ptolemaic' cartography. The Alexandrine concept that the fundamental principle of geography was the accurate fixing of position, if possible, by latitude and longitude astronomically determined, made a deep impression upon many fifteenth-century map-makers. By 1475 maps accompanying manuscript copies of the *Geography* were beginning to include maps drawn on the new principle.[2] Further evidence of its inspiriting effect comes from Fontana's work. Stirred up by his reading of Ptolemy, Fontana inquires what is the character of the lands beyond the Indian Ocean. As a result of investigation he is able to assure us that recent cosmographers, and

especially those who owe their information to true experience and the fruits of wide travel and diligent navigation, have found beyond the equinoctial circle on its southern side a notable habitable

[1] The word *Almagest* is derived from the Arabic *Al-Majisti*, or *Al-Mijisti* which is itself derived from the Greek *megistos*, i.e. greatest.

[2] *Vide* E. G. R. Taylor: *Tudor Geography*, p. 141.

land uncovered by water together with many famous islands. . . . Nor indeed is any of this area between the Torrid Zone and the extreme South uninhabitable, for from these islands and lands there came to us precious articles of almost every kind including beautiful pearls and silks.[1]

Now, if we assume that this passage is no later interpolation, but a *bonâ fide* statement dating from the fifteenth century, then so far from implying that the Polan tradition was moribund, it suggests that evidences of the great wealth of the tropical civilization of Asia were still forthcoming to whet the appetites of the Mediterranean merchants and to spur them on to new maritime endeavour. It would seem, therefore, that the East still lived in the imagination of the West—the works of Pierre d'Ailly and Pope Pius notwithstanding. One of the most important things about this revival (or should it be survival ?) of belief in the existence of habitable land south of the Equator is the fact that, in Fontana's case at any rate, it was held in conjunction with the opinion that more than half of the earth's surface was under water. Exactly why Fontana believed this is explained for us in a little treatise called *Libellus de Speculo Mikesi.*[2] Affirming that the earth, as the heavier element, should be enclosed by the sphere of water, he proceeds to argue that God was none the less able to produce a contrary state of equilibrium and that naturally, not miraculously, for the following rather obscure reason. The earth's centre of gravity coincides with the centre of the empyrean (which, needless to say, is regarded as geocentric), but the centre of its magnitude does not, presumably because

[1] *op. cit.:* fol. 90 r. 'Sed qui successerunt cosmographi et proprie qui vera experientia et peragratione itinerum et diligenti navigatione certiores facti sunt invenerunt ultra circulum equinoctiali suppositum versus Austrum esse partem notabilem habitabilem ab aqua discoperta(m) et insulas multas atque famosas . . . Nec est omnino illa zona inter torridam et extremam Australem inhabitabilis ex quibus insulis et partibus ad nos deferentur optimata omnis fere generis et margaritae gemmae pulcherrimae et sericum et multa pretiosa.'

[2] *Vide* Lynn Thorndike: *History of Magic and Experimental Science,* Vol. IV, pp. 175–6.

in one part of the globe the waters were commonly supposed to be piled up like mountains—under the influence of stellar attraction—rising higher towards the sky than the tops of the highest terrestrial mountains.[1] Consequently, a certain amount of land—a fourth according to the popular belief[2]—is uncovered by water. Now, according to Fontana's information, this habitable quarter of the globe is partly in the northern and partly in the southern hemisphere. Therefore it is certain, he avers, that each of the zones containing land is covered with water—and so uninhabitable—for at least half its longitude. This half, clearly, was the western half.

This is a most significant assertion, as Professor Lynn Thorndike has recently pointed out,[3] for it shows that the relation of the Aristotelian thesis of heavy and light as applied to the spheres of earth and water has not been fully realized in connexion with the discovery of lands in the western hemisphere. From what Fontana tells us it is evident that the Venetians of the mid-fifteenth century possessed considerable knowledge of the Indian Ocean and lands adjacent to it south of the Equator. But the acceptance of such information would make it difficult, if not impossible, for men holding the general belief that earth was heavier than water and that the *oikoumenē* was surrounded by the Ocean, to hold to the likelihood of there being any new landmass in the western hemisphere. Such men, if they were thoroughly convinced, could only seek to dissuade their fellow Italians from westward enterprises, whether promoted for the purpose of finding Cathay or a new world. How far this belief was held we have no means of knowing. It may well be that the Portuguese, on finding that the west coast of Africa extended south of the Equator, inclined to such a view. Be this as it may, it is almost past argument that the advocates of the western route to the 'Indies' were of one mind in supposing that there was no sizeable area of land between 'the extremity of Spain in the

[1] Cf. Ristoro of Arezzo: *La composizione del mondi di Ristoro* . . ., VI, 5, vii.

[2] *Vide ante*, chap. vii.

[3] *op. cit.*: Vol. IV, p. 162.

west and the beginning of India in the east'.[1] Even after such a landmass had actually been found, the world was reluctant to accept it as such. Columbus was not the only explorer who died clinging to the conviction. Unfortunately for the validity of the theory, we shall look in vain in the journals and letters of the American pioneers for any allusion to Aristotelian physical theory.

At the same time, Fontana's contention, based, as it was, on the belief in a 'notable habitable land', abounding in mineral wealth, in equatorial latitudes, coupled with the even more popular belief that Paradise was situated somewhere, may help us to explain Columbus' strong inclination to sail south-westwards, rather than due west. This inclination was particularly manifest after his first voyage[2] and is revealed, among other ways, by his under-scoring of every statement in d'Ailly's *Imago Mundi* relating to the temperate quality of the equatorial regions and to their economic resources. In one instance, he adds, with almost palpable satisfaction, that the discoveries of the Portuguese have proved the truth of the conjectures of earlier writers.[3] Furthermore, he accepts the view that Paradise lies towards the south, rather than towards the east.[4]

The same mingling of old myth and new knowledge, the

[1] On the lure of the Atlantic Islands, real and legendary, for the fifteenth-century explorer, *vide* W. H. Babcock: *Legendary Islands of the Atlantic*; G. E. Nunn: *The Geographical Conceptions of Columbus* (*passim*); J. Cortesão: 'The Pre-Columbian Discovery of America', *Geographical Journal*, Jan. 1937, and G. R. Crone: 'The Alleged Pre-Columbian Discovery of America', *Geographical Journal*, May 1937.

[2] *Vide* Cecil Jane in the Introduction to Vol. I., pp. cxiv *et seq.*, of *Voyages of Columbus* (Hakluyt Society).

[3] *Ibid.*: p. cxv.

[4] *Ibid.* In as far as the early views of Columbus and his companions are known, it is unlikely that they were very profound, if, indeed, they were reasonable. The much-advertised Columbian annotations of d'Ailly's *Imago Mundi* appear to have been made *after* the first voyage (*vide* C. Jane: *op. cit.*, Vol. I, pp. xc *et seq.*). On the other hand, there is no evidence of any fundamental change in his ideas. Judging from the scanty evidences, the views which he entertained at the time of his last voyage were essentially the same as those held at the time of his first voyage (*Ibid.*: p. cxvi).

same clinging to outworn traditions and unscientific method, characterize the geographical literature of the closing years of the fifteenth, and even of the early years of the sixteenth century. For when the Italian Renaissance was at its height, John Michael Albert, a physician practising in Bergamo (1460–90)—an educated man, therefore, and one who possessed at least the average geographical knowledge of his day—wrote that the Torrid Zone was 'arid, sterile and uninhabited',[1] notwithstanding that the Equator had been crossed more than ten years previously and the Tropic of Cancer more than forty. Further, he maintains that Ristoro of Arezzo (whom he follows closely) was right in arguing against the existence of the Antipodes, or, for that matter, land at all in the southern hemisphere, firstly, on the astrological ground that there are more stars in the northern hemisphere, and that the heads of the animals in the signs of the Zodiac all point northwards, their feet pointing southwards in every case; and, secondly, on the ground that if there were land far to the south 'men, surely, would have navigated it and crossed over to this southern land, and have learned the customs of those parts. But this has not happened, therefore those parts are entirely covered by water.'[2]

Alessandro Achillini, who wrote a treatise on the elements in 1505, is equally reluctant to abandon the legacy of the past, for he discusses the question whether the torrid zone is habitable in much the same way as Aristotle had done, and without any allusion to recent explorations. 'It is unprofitable', he maintains, 'to ask whether fruits grow there or whether the air there is most temperate, or whether the animals living there have temperate constitutions, or whether the Terrestrial Paradise is there', for these are 'matters which natural experience does not reveal to us'.[3]

On the other hand, in an exactly contemporaneous work, John of Glogau, who, as a professor in Cracow University, was geographically remote from the centre of maritime

[1] *De Constitutione Mundi*: III, 4, written *c.* 1483–90.

[2] *Ibid.*: VIII, 8.

[3] *De Elementis*: III, 19.

activity, wrote that the existence of the rich and populous island of Ceylon clearly refuted belief in an uninhabitable torrid zone. Moreover, he also knew that the Portuguese had discovered a new world in equatorial latitudes and that geographers had shown the so-called frigid zone to the north of the Arctic Circle to be inhabited.[1]

With the publication of the *Itinerarium Portugallensium* by Archangelo Madrignano, *c.* 1508, and the *Geography* of Dominicus Marius Niger, a few years later, the new age may fairly be said to have been ushered in, but many a year was to elapse before all the geographical ideas of the Renaissance had passed away.[2]

In the last analysis, however, to try to reconstruct the geographical ideas of the Renaissance is, as we have recently been reminded,

> to feign a shape for what was shapeless. . . . On the whole, the men of the Renaissance were too busy exhuming the ideas of the Ancients to have any of their own, and the ideas exhumed had little unity among themselves, the simple reason being that the ancients were of various races and many centuries, not a single homogeneous school. . . . Europe was just a whispering-gallery, echoing every voice that revived to speak in it.[3]

Babel though it was, the wealth of information, rumour and suggestion which travelled round its walls succeeded in broadening man's intellectual outlook, and so prepared him for the era of constructive thinking and synthesis about to dawn.

[1] *Introductorium Compendiosum in Tractatum Sphere Materialis Magistri Joannis de Sacrobusto . . .*: cap. 2.

[2] Thus, for instance, men continued to believe that the sea was surrounded on all sides by land, and consequently, that the surface of the land was not less than the sea, but perhaps of greater extent. Indeed, this was the prevailing opinion in the sixteenth and seventeenth centuries and was only definitely dispelled by the voyages of Captain Cook in the eighteenth century.

[3] T. E. Jessop in *France* . . . (edited by R. L. Graeme Ritchie): pp. 26–7.

CHAPTER X

THE BEGINNINGS OF GEOGRAPHICAL SCIENCE

THROUGHOUT the Middle Ages the open ocean inspired a great fear in mariners, not only because of the dangers inherent in it, but also because they were unable to fix with any certainty the position of their ships upon its surface. This inability had a very crippling effect upon seafaring, although, of course, it did not prevent men, here and there, from carrying out remarkable feats of exploration. These feats appear all the more remarkable when we recall the fact that, up to the thirteenth century, sailors had little more than the Pole Star and their own wits to guide them, and that their boats were no bigger than barges and often no more seaworthy. Under these circumstances, the wonder of it is not so much that they succeeded in making any discoveries—for many a man has been driven out of his course on to strange shores—but rather that they should have ever returned alive to tell the tale.

So long as maritime enterprise was largely confined to the Mediterranean region, where there was seldom need to be out of sight of land for more than a few hours at a time, sailing ' by instinct ' seems to have served these hardy pioneers well enough. But when in the early years of the fifteenth century men began to plan the exploration of the unknown equatorial and southern oceans, it became quite clear that empiricism in nautical matters would no longer suffice. The rise of nautical science was the logical outcome of the needs created by these expeditions.

What was the state of navigation at the dawn of this new

era? What, for instance, were the aids commonly employed, and how did men pilot their vessels over uncharted waters?

Undoubtedly, the instrument to which they owed most was the compass, the earliest references to which, in Christian literature, occur in the early twelfth-century works of Alexander Neckam [1] and Jacques de Vitry.[2] In view of the belief that the compass was used long before this time by the Arabs and the Chinese [3] for navigation purposes, it is, at first sight, curious that the earliest Persian and Arabic [4] references to it are somewhat later than the Latin. It is explainable if we realize that such a discovery would not be published by the pilots who presumably made it, nor by their associates who had every reason for keeping it dark; if it were published at all, this would be done—and only later—by outsiders. The fact that Neckam's mention of the compass is quite casual—it reads as follows: 'When the mariners cannot see the sun clearly in murky weather, or at night, and cannot tell which way their prow is tending, they put a needle above a magnet which revolves until its point looks North and then stands still' [5]—suggests that his readers are expected to understand the import of it and presupposes that it had been in existence for some long time prior to the time of his writing. It is just possible that one reason why the compass was not in common use in Christendom until the fourteenth century was that those in possession of the secret of it were afraid to reveal it lest they should be suspected of magic or that sailors were equally afraid to employ the new device for the same reason! [6] The documentary evidence warranting such a view is, how-

[1] *De Naturis Rerum.*

[2] *Libri duo . . . Occidentalis Historiae.*

[3] *Vide Encyclopaedia Britannica,* 11th Edition. Article: Compass.

[4] The first reference in a Moslem work is found in a Persian collection of anecdotes written by Al-'Awfī *c.* 1230. One story in the collection tells how the author as a sailor found his way by means of a 'fish' rubbed with a magnet.

[5] *De Naturis Rerum:* II, 98.

[6] *Vide* Lynn Thorndike: *History of Magic and Experimental Science,* Vol. II, p. 388.

ever, very slight.[1] By the fourteenth century, the compass was being employed not only for nautical purposes but also for terrestrial. The craftsmen of Nuremburg, far-famed, then as now, for their instruments of precision, invented a new kind of compass which they styled the *organum viatorum*. This incorporated a miniature sundial into the compass mechanism and so combined the functions of a direction finder with that of a time-piece. With the introduction, towards the end of the fifteenth century, of road maps which required careful orientation, this instrument became an indispensable part of any land traveller's equipment.

Two other instruments were in fairly general usage by the fourteenth century, namely, the cross-staff and the astrolabe. The former appears to have been employed by astronomers at a very early date, though there is no definite evidence that seamen made a regular practice of using it before the twelfth or thirteenth century. In its simplest form it consisted of two light wooden battens of unequal length. The longer of the two—some three feet in length—was the 'staff' and was graduated in such a way that when the 'cross' was in position it recorded the Pole Star's altitude at sight. The 'cross' was made so that it could slide upon the staff at right angles; its length was normally about 26 inches and it was fitted with

[1] In an article recently published in the *Mariner's Mirror* (Jan. 1937) Dr. Heinrich Winter presses the claims of the Vikings to have invented the compass (independently of the Chinese). Using the facts that several of their early Sagas (those of the Faroe Isles in particular) mention the lodestone floating in a container, that they were in the habit of determining longitude by the help of a water-clock and latitude by means of an instrument of the *gnomon* type, that they were making *ocean* voyages before the year 1000, that from the ninth century onwards they penetrated into the Mediterranean and there, with continuous reinforcements of men and ships from home, founded kingdoms of their own—in particular, the Kingdom of Calabria and Apulia—that Roger II established a university at Salerno in 1100 and in addition, had a map of the world made to his order; he argues that in face of this accumulated evidence, it is not merely a possibility, but rather a strong probability that it was the Vikings who introduced the pristine forms of the compass to the Italians.—Vol. XXIII, p. 102.

sights or 'pinules'. A sight was also fixed on the end of the 'staff' for the eye to see, at one and the same time, the two objects—commonly the Pole Star and the horizon—whose angular distance apart could then be measured.[1]

More convenient than the cross-staff for taking altitudes was the astrolabe. As a purely astronomical instrument it dates from the days of Eratosthenes and Hipparchus, but it does not seem to have been used for nautical observations, in Europe at any rate, until the Moslem era. The 'standard' model was usually made of brass or copper (weight being considered desirable in order to steady it when in use, especially at sea) and varied in diameter from about 2 inches to 1 foot and more. The observational part consisted of a centrally pivoted alidade about 1½ inches with two sight vanes and pinhole sights which could be rotated round a quadrant of degrees usually numbered from 0° at the horizontal to 90° at the zenith.[2] The verticality of the instrument was assured by the symmetry of the two sides and the suitable construction of the ring by which it was hung. The alidade moved upon a centre the size of a goose quill. When the instrument was suspended, the pointer was directed by hand to the object, and the angle (i.e. the Pole Star's altitude) read off from the graduated quadrant. Naturally the movement of the ship made it very difficult to attain any great degree of accuracy, while the fact that, ideally, three observers were necessary to make each calculation—one to hold the instrument, another to measure the altitude and a third to read it off—merely served to increase the liability to error.

Mainly because of these limitations, attempts were con-

[1] The sun was not used until the late fifteenth century, as the calculation of latitude from it required a table of declinations. *Vide infra*, pp. 229–31.

[2] In addition, many astrolabes carried a circular calendar scale (for finding the sun's place in the Zodiac on any given day and month) a circular rotating star-map (by means of this the daily movement of the sun and principal stars as they rise and set could be demonstrated) and a plate engraved with a polar projection of the circles of altitude of the celestial sphere. *Vide* R. T. Gunther: 'The Astrolabe: Its Uses and Derivations, *Scottish Geographical Mag.*, Vol. XLIII, p. 136.

tinually being made, especially by the Arabs, to improve upon the 'standard' model. Al-Zarqālī, the early twelfth-century astronomer, was the leading experimenter here. His improved model became popular, not only in Arabic circles, but, after his description of it had been translated into Latin (1263), in Christendom also. One result of this was that most Christian scholars and astronomers were content to accept the findings of their Arabic contemporaries. However, at least two Englishmen showed a more independent spirit in this matter; they were Adelard of Bath, who wrote an original treatise on the astrolabe in 1144, and Robertus Anglicus, who in the following century (1276) adapted a simple Arab form of the instrument, which he called the *quadrans*, i.e. quadrant. His description of this—*Tractatus quadrantis*—soon became popular, being translated into several languages. A few years later, a Jew, Jacob ben Machir, described a new and better *quadrans*.[1] From then on, the few who knew called Robert's instrument *quadrans vetus*, and Jacob's *quadrans novus*. For the Middle Ages, when the rhythm of life was incomparably slower than it is now, this was rather a quick change and shows a growing interest in experimental science and observation. But neither instrument appears to have been in regular use in Portugal before the time of the Navigator.[2]

Supplementing the practical equipment of sailors in the late Middle Ages were the *portolani* or 'reading-maps', the lineal descendants of the ancient *peripli*. These incorporated the

[1] *Vide Quadrans Judaicus* or *Quadrans novus*, Bibl. Nat., Paris, Latin MS. 7,437. The earliest quadrants merely consisted of a quarter (hence the name) of the astrolabe enlarged, with fixed sights placed in, or parallel to, the vertical line passing through zero. A fine thread or wire stretched by a plummet, indicated the altitude on the graduated limb. The chief advantage of this instrument was the increased radius which promised to give more accurate readings than an astrolabe of the *same over-all size*. The absence of accurate degree subdivisions, the difficulty of sighting on to a star from the deck of a rolling ship and the thickness of the wire, were impediments to accuracy that did not readily admit of remedy, however.

[2] *Vide* J. Bensaude: *L'Astronomie nautique au Portugal a l'époque des Grandes Découvertes*, p. 35.

cumulative experience of centuries of observations when the lode-line was the sole navigating instrument other than the shipmaster's eye. Compiled by seafaring men for themselves, they had only a limited circulation and seldom found their way into libraries and archives. Most of them were used until they wore out with age, and few of them, if any, have survived in their *original* form. But there are plenty of allusions to them and, occasionally, it is obvious that the writers had a copy of a book of sailing directions in front of them. Marino Sanuto, writing *c.* 1306–20, even went to the trouble of copying such a work for insertion into his *Liber secretorum fidelium crucis*.[1] Even before his time, however, we encounter allusions to sailing directions, and quotations from them. Thus, the anonymous work, entitled *Gesta regis Ricardi*, and describing the exploits of Richard Lionheart, gives much rutter-like information concerning the routes followed by him, while Roger Hoveden, in his Chronicle of England (*c.* 1200), describes Richard's itinerary in a similar fashion.[2] There is, moreover, a remarkable Scandinavian sailing direction, dating from the late thirteenth century, which gives the route between Denmark and Syria. The existence of several copies of this document suggests that the people of North-Western Europe were not unfamiliar with navigation manuals at this epoch. Then again, there is the German 'Sea-book'

[1] II, 25–6.

[2] *Chronica* . . . (Rolls Series, edited by W. Stubbs): Vol. III, pp. 39 *et seq*. 'Deinde venit rex ad civitatem quae dicitur Tharracene, ubi quondam erat portus cupro tectus.

'Deinde venit ipse ad Garillam: et inde ad castellum quod dicitur le Cap del Espurun. Hic est divisio terrae Romanorum et terrae regis Siciliae, in illa parte quae dicitur principatus capuae.

'Deinde praeteriit rex insulam quae dicitur Lapantee quae distat a Gaeta civitate per quadraginta milliaria. Pantee terra fuit Pilati: et ibi erat portus cupro tectus.

'Deinde praeteriit rex insulam quae dicitur Istele major, quae semper fumat. Haec insula dicitur accensa fuisse ab insula quae dicitur Vulcanus: igne volante, et comburente, ut dicitur, mare et pisces. Ab hac non multum distat insula quae dicitur Isle de Girun, ubi est castellum bonum et portus. . . .'

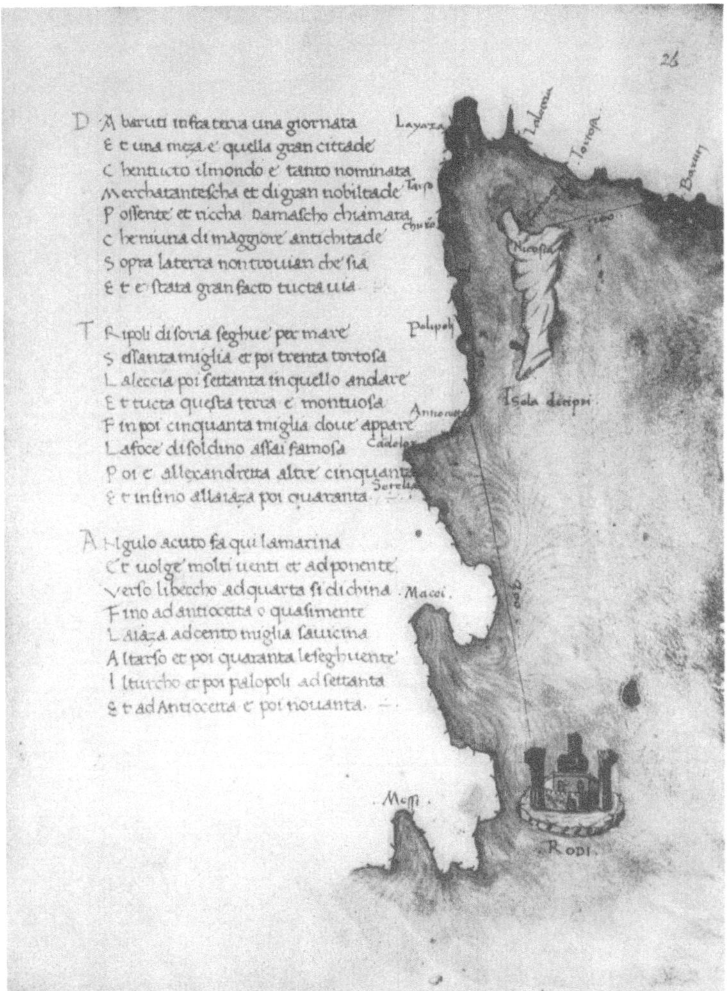

XIX

MAP OF LEVANTINE COASTS FROM DATI'S *La Sfera*
(*Brit. Mus., Add. MSS.* 24,942)

of the fifteenth century (but incorporating, according to Nordenskiold,[1] materials at least 100 years old), giving a description of the Ocean coasts of Europe in the Baltic Sea. Nor must we forget the remarkable rhyming cosmographical treatise of the fourteenth century, of Florentine provenance, known (by the title of its first section) as *La Sfera*, and usually ascribed to Leonardo Dati.[2] The second section dealing with the rules for navigation and the determination of position at sea, and containing, in the margins, maps of the various Mediterranean coasts, though naturally very crude at this date, is obviously based upon some existing manual; whether the author supposed that the work itself would ever become a text-book for sailors or not, we cannot say. The only copies that have come down to us are elaborately illuminated examples and were clearly prepared for presentation purposes.[3]

Now if Dati's treatise is at all representative of a class of literature, and not an isolated and unique document, then it is not unreasonable to suppose that the earliest Portolan charts formed the illustrations for the manuals. In this case, then the fact already observed, namely, the polished and stylized ' finish ' of the earliest-known charts, offers a presumption that the manuals they illustrated were by no means a recent innovation. As far as Portugal is concerned, however, it is only fair to say that neither Zurara nor any other fifteenth-century historian of that country alludes to the existence of such documents.[4] Sailing charts, on the other hand, come in for considerable notice.[5]

THE PORTUGUESE CONTRIBUTION

Nowadays it is widely accepted that the Portuguese were the founders of nautical science. Before their time maritime

[1] *Vide Periplus:* pp. 103–4. [2] *Ibid.:* p. 45.

[3] *Vide* Add. MSS. 22,329, 24,942 (British Museum) and Plate No. XIX.

[4] The only English sailing direction of the Middle Ages which appears to have survived has been published by the Hakluyt Society under the title *Sailing Directions for the Circumnavigation of England . . . from a 15th-century MS.* Edited by J. Gairdner, 1889.

[5] *Vide ante,* chap. viii, p. 199.

enterprise had been conducted largely by private individuals or by small companies of adventurers. As few of these men recorded their voyages, kept logs or drew charts, their experiences had little cumulative value and each generation remained almost as ignorant as its predecessor. At the outset of his undertaking Prince Henry realized that this wastage of effort and knowledge must cease before there could be any substantial advance in the science of navigation. As a second pre-requisite of scientific progress, he realized also that there must be a more effective collaboration of theory and practice. To ensure these two ends and, at the same time, to promote positive nautical research, he founded the Academy, or School of Sagres.[1] Here came scholar and sailor, as well as many a distinguished mathematician and astronomer. Most famous of all the men to be attracted thither was the Catalan cartographer, Jacomo of Majorca, 'a man very learned in the art of navigation and in the construction of charts and instruments, whom,' so Barros tells us, 'it was very desirable to bring to this kingdom to enable Portuguese officers to learn from that master'.[2] Henry also secured the active co-operation of other Catalans, likewise of Italians and Jews, in collecting accounts of medieval exploration and in the study of current astronomical problems. In this work the Navigator enjoyed the help of his brother Pedro who, on one occasion, brought home from Italy a number of maps 'by which he [i.e. Henry] was much helped and furthered in his discoveries'.[3] To seal his interest in nautical matters, he established a Chair of Astronomy in 1431 at the University of Lisbon.

While little of the work of the Sagres School has survived the centuries—we know nothing of Jacomo's own work there—it is evident that its achievements were none the less solid and enduring. Proof of this is forthcoming in at least two directions: first, in a change of intellectual outlook. Gradually experience, rather than tradition, became the guiding principle of Portuguese thought—a principle that is commonly

[1] *c.* 1416. [2] *Asia*: Decade I, Book I, chap. 16.
[3] A. Galvão: *Discoveries of the World*: p. 67 (Hakluyt Society).

expressed in such words as these: 'Experience disabuses us of the errors and fictions . . . of ancient cosmographers.'[1] In the second place, as we have already observed, things began to happen about this time. The age-long spell cast by fear of the unknown was broken almost overnight, while the fabulous notions surrounding the Torrid Zone were fast being dispelled. Each year saw new territory being added to the habitable earth in latitudes which had hitherto been deemed uninhabitable.

As this exploration progressed it became increasingly clear that the existing stock-in-trade of the mariner was inadequate, particularly in regard to the calculation of latitudes in equatorial waters. It was simple enough to find the latitude of a place in the temperate regions of the Northern Hemisphere—provided the night was clear—by observing the altitude of the Stella Maris or Tramontana. As sailors drew near to the Equator this method ceased to be practicable, for the elevations became so low that they could not be measured by the astrolabe to within 3° or 4° of the actual figure. As early as 1456 Ca'da Mosto observed that the Pole Star appeared to him, in the Rio Grande[2] region, to be so low in the sky that he could barely obtain his position from it.[3] When, a few years later,[4] the Equator was crossed for the first time by a European, a new guide was imperative.

King John II's Junta of astronomers and mathematicians solved the problem by utilizing the relation between latitude, the sun's noon altitude and its declination. To facilitate the sailor's job of calculating this, tables of declination were prepared. These, in all probability, were derived from the *Almanach Perpetuum*[5] of Abraham Zacuto, the professor of

[1] Duarte Pacheco: *op. cit.*, Book II, chap. 11. [2] i.e. River Jeba, 12° N.

[3] *Vide Relation des voyages à la côte occidentale d'Afrique d'Alvise de Ca'da Moste*, 1455–7 (translated by J. Temporal, 1895): pp. 178–9.

[4] *c.* 1471.

[5] That the equally famous contemporary *Ephemerides* of the German astronomer Regiomontanus were not used (as has been confidently affirmed by Alexander von Humboldt and others) may be deduced from the fact that the value of the ecliptic adopted by the Portuguese is that of the *Almanach* and not of the *Ephemerides*.

astronomy at Salamanca, *c.* 1473, and the chief representative of the Jewish astronomical school of his day. Being somewhat too technical for the ordinary shipmaster, a more practical manual, entitled *Regimento do Astrolabio e do Quadrante*, was compiled from it. By the proper use of the astrolabe, or quadrant, and these tables, mariners could ascertain their approximate position with regard to the Equator. Judging from the opinion of Master John, pilot to Cabral on his expedition of 1500, there was certainly more to be said for the new method than the old.

> It seems to me [he says] almost impossible to take the height of any star at sea, for no matter how much I labour at it or how little the ship rolls, there are mistakes of 4° to 5° so that it can only be done on land . . . At sea it is better to be guided by the height of the sun than by any star and it is better to use the astrolabe than the quadrant or any other instrument.[1]

But to take an accurate observation of the sun's noon altitude in low latitudes from the deck of a small sailing ship with a comparatively high centre of gravity, and therefore prone to pitch and roll, using a hand instrument with no telescopic sights and graduated to degrees only, must have been quite difficult enough. To get the latitude correct to within a degree or so was, in the circumstances, a notable achievement.

The method of arriving at these calculations deserves some attention, and once again we can do no better than turn to Pacheco's *Esmeraldo*. After a general preamble the author proceeds to formulate four rules:

1. When the sun is between the observer and the Equator, whether to the north or south of the Equator,
Latitude = 90°—(Altitude of sun at noon—its Declination).

2. When the observer is between the sun and the Equator, and the altitude and declination together equal more than 90°,
Latitude (N. or S.) = Altitude + Declination − 90°.

[1] *Vide Alguns documentos do archivo nacional da Torre do Tombo*, Lisbon, 1892: pp. 121–3. This view was confirmed by the majority of Portuguese navigators.

3. When the observer is between the sun and Equator, and the altitude and the declination together equal less than 90°,

Latitude (N. or S.) = 90° − (Declination + Altitude).[1]

4. When the Equator is between the sun and the observer, as above.

Compared with the methods described in the Munich *Regimento*,[2] where seven cases are postulated, the directions in the *Esmeraldo* are remarkably succinct and comprehensive. In point of accuracy there is little to choose between the lists of latitudes given in the two works; if anything, Pacheco's is the more precise. Naturally the degree of accuracy is greatest in those parts of Europe and Africa where measurements had been taken for a comparatively long time; but even in southern latitudes Pacheco's figures are by no means wide of the mark. Thus he places the Cape of Good Hope in 34° 30′ S,[3] Sofala 20° S,[4] and Cape San Roque 3° 30′ S.[5] The enumeration of places with their latitudes south of the Equator is especially noteworthy when we recall that King Manuel issued a decree in 1504 ordering the suppression of all information, including latitudes, bearing on the navigation of the African coast beyond Rio do Padram, i.e. 7° S. If, as seems likely,[6] the *Esmeraldo* was banned by the national inquisitors, this may well have been one of the contributory reasons.

The problem of longitude was less tractable, and Pacheco, in common with his contemporaries, avoids it, notwithstanding an assurance to the contrary early in his work. He contents himself with the rather naïve statement that

degrees of longitude are counted from orient to occident, which mariners call east and west, and this is difficult to ascertain because

[1] A moment's reflection will suffice to show that Pacheco was in error here, for when the observer is between the sun and the Equator the sun's noon altitude and the declination must always be more than 90°. This formula, however, holds good for the case which he goes on to describe.

[2] A manual of practical astronomy compiled in Germany at the beginning of the sixteenth century.

[3] Actually 34° 22′ S. [4] Only 10′ out. [5] 2° out.

[6] *Vide* Introduction to Hakluyt Society Edition: p. xxix.

they have no firm and fixed point (of reference) as are the poles for latitude, and of this I will say no more.[1]

It was not until the following decade that the subject was first dealt with in the text-books of navigation, and then only in a tentative fashion. In his *Regimento* for longitude, *c.* 1517,[2] Ruy Faleiro proposed no less than three different methods: these involved the use of lunar latitudes,[3] the conjunctions and oppositions of the sun and moon, and the deviation of the compass. Francisco Faleiro, a few years later, reiterated the last of these arguments, without, however, attempting to establish it.[4] João de Castro and Pedro Nunes,[5] for all their astronomical learning, were able to make no positive contribution to the subject, but de Castro did at least come to the conclusion that 'magnetic deviation has no connexion with differences of meridian'.[6] And so the search proceeded until the goal was reached, some two centuries later, as a result of the invention of the sextant and chronometer. Until then the general practice of calculating 'the height of east-west'—as the Portuguese paraphrased the word 'longitude'—is admirably explained in the following sixteenth (?) century description:

There be some that are very inquisitive to have a way to get the longitude, but that is too tedious for seamen, since it requireth the deep knowledge of astronomy. Wherefore I would not have

[1] *op. cit.*: I, 8.

[2] Known only from an extract in Pigafetta's *Replazione del primo viaggio intorno al mondo*; vide pp. 39–40 ('Alpes' edition, 1928).

[3] Regiomontanus, the German astronomer, had advocated this method in his *Ephemerides* (1474). It consisted of measuring the angular distance between the moon and a given star, but it was of only very limited value at this time owing to the imperfect knowledge of the true positions of the moon and stars and the lack of instrumental means by which such distances could be accurately determined.

[4] *Vide Arte del Marear*: cap. VIII (1532).

[5] *Tratado em defensam da carta de marear*: (1537) pp. 140 *et seq.* J. Bensaude's facsimile edition, 1915.

[6] *Roteiro de Lisboa a Goa*: (1538) pp. 384 *et seq.* Edition of 1882.

any man think that the longitude is to be found at sea by any instrument: so let no seamen trouble themselves with any such rule, but (according to their accustomed manner) let them keep a perfect account and reckoning of the way of their ship.[1]

Such record of 'the way of the ship' appears to have been little more than a 'dead-reckoning' of the daily run, marked in chalk on a wooden board, which folded like a book and from which, each day, a position for the ship was deduced. The usual manner of arriving at this daily figure was the so-called 'Dutchman's log' method; it consisted simply in throwing into the water, from the bows of the ship, something that would float (spittle, more often than not !), and noting with the sand-glass the time interval between its drifting past two observers standing on the deck at a known distance apart. No other method is mentioned—except guessing—before 1559.[2] Needless to say, accuracy of results was not to be hoped for from such rough-and-ready means, especially when, to these errors, we add those accruing from the under- (and over-) estimation of the influence of winds, tides and currents upon the ship's speed. As the results were frequently so bad as to be palpably wrong, it is not surprising to find that many mariners preferred to 'gesse whereabouts they be touching the degrees of longitude'.[3]

Second only in importance to the calculation of latitude and longitude was the computation of the ebb and flow of the tide, for, as Pacheco reminds us, 'with this knowledge . . . ships will be able to enter safely rivers and other places for

[1] Quoted in *Encyclopaedia Britannica*, 11th Edition. Article: *Navigation.*

[2] The use of 'a parfait clocke' was suggested in that year by W. Cunningham in his *Cosmographical Glasse*, *vide* E. G. R. Taylor: *Tudor Geography*, pp. 26–7.

[3] Thomas Stevens, writing from Goa in 1579, *vide* Hakluyt's *Voyages:* Vol. XI (Goldsmid's edition), p. 266. On the related question of the prevailing ideas concerning the length of a geographical degree, *vide* G. E. Nunn: *The Geographical Conceptions of Columbus*; *Ibid.*: 'Marinus of Tyre's Place in the Columbus Concepts' (*Imago Mundi*, Vol. II, 1937) and E. G. R. Taylor: 'Some Notes on Early Ideas of the Form and Size of the Earth' (*Geographical Journal*, Vol. LXXXV, 1935).

which it is necessary to know the state of the tide'.[1] Observations upon tidal behaviour antedate the days of the Portuguese pioneers,[2] but it is not until the appearance of Pacheco's *Esmeraldo* . . . *c.* 1505, that we get the first systematic and practical discussion of the subject. In two introductory chapters [3] he describes how it is possible 'to compute the ebb and flow of the sea in the greater part of Spain and likewise in other regions where there are tides'. The relation between the moon's course and the tides he takes for granted, but he grossly underestimates the complexity of the subject when he declares that a knowledge of the lunar cycle will enable a mariner to judge how the tide stands, no matter where he is. He did not know, what is now common knowledge, that the constituents of a tidal régime vary from place to place, and he seldom seemed to notice the varying effect of coastal configuration on the rhythm and occurrence of tides.[4] How little he comprehended the implications even of his own observations is clearly seen where he considers the relationship between the lunar cycle, solar time and the state of the tide. Here he starts from the fact that along the Portuguese coasts, high tide occurs at conjunction when the moon (and the sun, therefore) is in the north-east, i.e. at 3 a.m. 24 hours and 45 minutes later the moon is again in the north-east, but the sun will be a 'quarter point' farther east, i.e. north-east by east,[5] and so high tide will occur at 3.45 a.m. On the follow-

[1] *op. cit.*: I, 11. [2] *Vide ante*, chap. vii. [3] Chaps. 11, 12.

[4] In this respect, Pacheco lagged behind his contemporaries somewhat. For instance, Robert of York (*vide ante*, chap. ix) observed, in connexion with tidal rivers, that a tide which is manifest in the ocean when the moon is due south will not reach 30 miles upstream until the moon is SW. or thereabouts (*op. cit.*), while the author of the Catalan Atlas of 1375 speaks of 'the establishment of the port', i.e. the interval between the moon's meridian and the occurrence of the next high tide. As early as the twelfth century, Giraldus Cambrensis, as we have already seen (chap. vii), had noted that the difference in the hour of high water around the coasts of the Irish Sea was something more than a function of the moon's meridian.

[5] The sun, theoretically, goes round the eight points of the compass every 24 hours: therefore, it passes from one 'quarter point' to the next every 45 minutes.

ing day the moon will be in the north-east at 4.30 a.m., i.e. when the sun is east north-east. In other words, high tide comes a 'quarter point' later ever day. Accordingly, Pacheco says,

> he who wishes to calculate the tide must see in what point of the compass the sun is and then count the number of days which have passed since the conjunction, counting a quarter point of the compass for each 24 hours up to 15 days or less, and according to the position of the moon will be the tide.[1]

Up to this point his directions are explicit enough, but when he comes to the question of distributing the 45 minutes' daily time-lag round the points of the compass, he seems unable to grasp the implications of his own argument. Thus, having stated that, when the sun and moon are in conjunction, low water occurs off the coast of Spain at 9 a.m. (when, according to his reckoning, the sun and moon are in the south-east), he goes on to tell us that high tide will occur when the sun is in the south-west, i.e. at 3 p.m., and the next low tide when the sun is in the north-west, i.e. at 9 p.m., whereas, by hypothesis, high tide will occur at 3.11¼ p.m. and low tide at 9.22½ p.m. Apparently, he did not realize that anything was wrong until he had worked round the compass points to the east, by which time, presumably, he could see that his method of reckoning was going to give 9 a.m. for the next low tide, instead of 9.45 p.m., for at this juncture he interpolates a sentence to the effect that, as the moon is 45 minutes behind the sun, it will no longer make the tide at the same time as on the previous day. It is not until he comes to deal with the state of the tide when the sun is south-east by east that he assumes the 45 minutes' time-lag to be in operation.[2] However, apart from a few isolated lapses of this sort, Pacheco's treatment of nautical matters is highly meritorious.

The calculation of bearings and distances suffered from another limitation at this time, namely, the mariner's ignorance of the magnetic factor. Exactly when the Portuguese, or any

[1] I, 12. [2] *Loc. cit.*

of the other sixteenth-century naval peoples, first began to be aware of it, it is impossible to say. The fact that the North point of a compass does not normally point to the true pole appears to have been noticed at an early date,[1] but there is little or no evidence to show that variations in this magnetic attraction were observed before the time of Columbus. At all events, the first authentic reference to this phenomenon is found in the Columbian Journal of the voyage of 1492[2]; however, nowhere in this journal is it implied that this was the first known observation, or even that it was unusual. The expressions used by Columbus suggest that the phenomenon was common knowledge in his day. Thus, he takes it for granted that his readers will know what is meant by 'needle nordesteau or noruesteau'. Martin Behaim alludes to the same thing on his globe—also of 1492—where he tells us: 'Those who navigate in this sea (= Southern Atlantic) should determine their course by the help of the astrolabe, for the compass does not function there.' As his information for this part of the globe was based exclusively upon Portuguese sources[3]—setting aside the possibility that some of his 'facts' may have been the product of an over-active imagination—it is obvious that men like Bartholomew Dias were aware of this deviation. At the same time it is equally obvious from the literature of the period that they did not know how serious it was, or how to correct it. A comparison between a modern map of West Africa and one based on the unadjusted plot of Pacheco's bearings and latitudes admirably illustrates the extent of the error.[4]

[1] Heinrich Winter has recently shown not only that the discovery of magnetic declination goes back at least to the thirteenth century, but that it was *rectified* as early as 1269. *Vide Imago Mundi*: Vol. II, p. 72. See also article by Dr. A. C. Mitchell in *Terrestrial Magnetism*, Sept. 1937.

[2] *The Journal of Columbus* (Hakluyt Society). Edited by C. R. Markham: pp. 24–5.

[3] *Vide* G. H. T. Kimble: 'Some Notes on Mediaeval Cartography'. . . . *Scottish Geographical Mag.*: Vol. XLIX, 1933, pp. 95 *et seq.*

[4] *Vide* Outline maps at end of *Esmeraldo* . . . (Hakluyt Society), constructed by W. F. Morris.

Although several of the great Portuguese astronomers alluded to the fact of this magnetic deviation, including the Faleiros and Pedro Nunes [1] (who did not attach much importance to the reports of pilots relating to it), it was not until almost the end of their golden age that any agreement was come to regarding the varying allowances to be made for this deflection, and then the only major correction generally accepted was a deviation of $11\frac{1}{4}°$ E in the Guinea Gulf.[2] João de Castro, almost alone of his contemporaries, realized, however, that this was a variable quantity when he says that 'at Lisbon the deviation is $7\frac{1}{2}°$ and the place where I find myself at this moment on the same meridian as Lisbon, the deviation is from 19° to 20°'.[3] from which he concludes, incidentally, that there can be no connexion between longitude and deviation.

Whatever the explanation of the dearth of navigation manuals in the fifteenth century—indeed, we might almost say their *absence*, the so-called *roteiro* of Vasco da Gama's first voyage being little more than a very simple diary, and very deficient in scientific observations—there is no doubt that the discovery of the Cape route to India at the end of the century gave a tremendous impetus to the detailed description of the new sea-route, for manuals on sailing, complete with details of sailing directions, began to appear in the first decade of the sixteenth century and were published thereafter at frequent intervals. At first these were the work of the Portuguese exclusively and were known as *roteiros*; similar works by Dutch, Spanish and English writers followed soon after. From these have sprung the indispensable *Pilot* books of to-day.

If we exclude some rather crude *roteiros* of uncertain date by Gonçalo Pires and João Rodrigues, then the earliest dated *rutter* [4] of the great age of discovery is Pacheco's *Esmeraldo*. In this connexion, as in the others, it deserves more than a passing reference, for, while there is reason to believe that

[1] *op. cit.*
[2] *Vide* E. G. R. Taylor: *Tudor Geography*, p. 68.
[3] *op. cit.*: p. 197.
[4] Anglicized form of *roteiro*.

it enjoyed only a short-lived fame,[1] it is none the less representative of the works that followed. Though the title is obscure, there is nothing obscure about the scope and aim of the book. In the author's own introductory words, his intention is to write 'a book of cosmography and navigation'. Because of the orientation of Portuguese interests he will limit himself to two of the four continents, Africa and Asia; all that concerns navigation he will describe at great length:

> Thus all the routes will be stated, giving the position of places and promontories in relation to one another, in order that the work may have an ordered basis and that the coast may be navigated in greater safety. Similarly, the land-marks and shallows, for these are essential; also the soundings, both their depths and the kind of bottom, whether it be of mud or sand or rocks or gravel or snags or shingle, together with the distance of the soundings from the land. Similarly the tides, whether they flow NE and SW as in our Spain, or north and south or east and west or NW and SE, which it is essential to know in order to enter and leave the bars and mouths of rivers; further, the altitudes of the poles, by which one may know the latitude of a place and its distance from the Equator.

He says he will also describe the inhabitants of the land of Ethiopia and their way of life.

Unfortunately, the project does not seem to have been entirely realized. Three of the five books contemplated are extant and a part of the fourth, carrying the description down to the Rio do Infante in South Africa. As far as it goes it certainly fulfils the author's promises. The coastwise indications, to take only one aspect of his work, are, on the whole, so precise that it has been found possible to identify almost every one of the several hundred land-marks described. The *Africa Pilot* can scarcely improve upon some of the descriptions.

In view of the growing practice of determining latitudes by the sun's noon altitude, it was only natural to find these coastwise descriptions prefaced by a chapter or so on this topic. Further, seeing that an elementary knowledge of cosmography

[1] *Vide* Introduction to Hakluyt Society Edition: p. xxviii *et seq.*

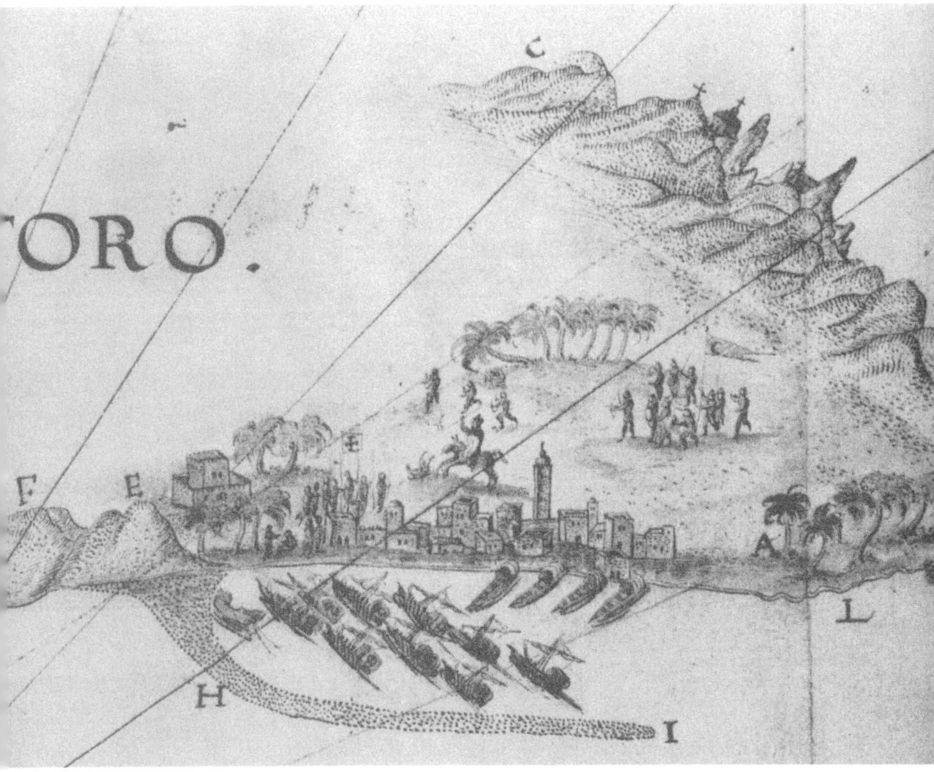

XX. TYPICAL ILLUSTRATION FROM THE *Roteiro* OF D. JOÃO DE CASTRO FROM INDIA TO SUEZ, 1541

(*Brit. Mus., Cotton MS. Tib. D. ix*)

was equally essential to seafarers, it was just as natural that the *Esmeraldo* and the later rutters should have another section on the terrestrial relationships of land and sea and similar topics.[1]

After the *Esmeraldo*, in point of time, come the *roteiros* of the Indian Ocean by João de Lisboa and André Pires. The work of the former (*c.* 1514) enjoyed a great reputation in its day and was probably a revised edition of an earlier *roteiro*, extended to include the coasts of India and Malay. As no Portuguese ship had then reached the islands of the East Indian Archipelago which are recorded in this work, it seems fairly certain that the author derived part of his information from contemporary Arab sailing directions. But the *Livro das Rotas*, as it was called, was not without its critics, one of whom, Diogo do Couto, the celebrated chronicler of India, sarcastically observed that he had little use for such pilots who were so clever at drawing charts and making great show with mathematical instruments, but who invariably ended by wrecking their ships on some shore and in losing their own lives.[2] Notable among other Portuguese *roteiros* are those of João de Castro: indeed it has been said that they mark a 'landmark in the history of nautical science',[3] the standard they set never being surpassed in Portugal. Little seems to have escaped the author's notice. Apart from his observations on magnetic deviation, already referred to, de Castro's descriptions of meteorological phenomena, such as the water-spout and halo, are extremely accurate, while his rulings on coastwise navigation leave nothing to be desired. The precision of his maps accompanying the descriptions of his voyages to Diu and the Red Sea, gives them the quality of hydrographic charts.[4]

In addition to these practical hand-books, the early decades

[1] By the middle of the sixteenth century it had become usual for the tables of latitudes and declinations to be added to this introductory material and also for pilot charts and sketches to illustrate the body of the work.

[2] *Dialogo do Soldado Practico Portuguez* (1790): chaps II, III.

[3] C. R. Boxer: 'Portuguese Roteiros, 1500–1700', *Mariner's Mirror*, Vol. XX, 1934, p. 176.

[4] *Vide* Plate No. XX.

of the sixteenth century gave us a number of works devoted to the more theoretical aspects of nautical science. The most important of these is the *Tratado em defensam da carta de Marear* of Pedro Nunes. The quarrels between Portugal and Spain over the possession of the Molucca Isles raised a whole series of scientific problems turning upon the detailed accuracy of the existing maps. In 1529, the year of the Congress of the Moluccas, Nunes was appointed cosmographer to the Portuguese Government. It was perfectly natural, therefore, that he should be preoccupied with the improvement of current cartographical technique. Unfortunately, he did not get very far in his quest, for although he considered many pertinent questions, almost his sole practical achievement was a revision of the existing tables of declination and dead-reckoning. On the theoretical side, his reputation rests mainly upon his advocacy of rigorously scientific cartography ; he was constantly stressing the need for solving the problems of portraying geographical co-ordinates and rhumb-lines on a plane surface. His great complaint was that maps lacking these data were 'worthless, although they have much gold and many flags, elephants and camels on them '.[1] In ways such as these Nunes prepared the ground for the Flemish pioneers of mathematical geography and particularly for the work of Mercator.

By an ironical chance, the publication of the *Tratado* . . . in 1537 coincided with the introduction of the Inquisition. It was not long before the hates and fears bred by this reign of terror began to infect the intellectual life of the Peninsula. In a world of heresy-hunters scientific investigations once again fared ill. Nautical science made little headway either in Portugal or Spain and although *roteiros* and sailing manuals continued to appear, they were, for the most part, simply new editions of the earlier works.

The scene quickly shifts from the Peninsula to the Low Countries. It is there that the era of scientific geography finally becomes established.

[1] Quoted by E. Prestage : *Portuguese Pioneers*, p. 325.

APPENDIX

DANTE'S GEOGRAPHICAL KNOWLEDGE

ALTHOUGH none of them are strictly geographical, Dante's works contain a large amount of geographical information and reveal, so we may believe, the kind of earth-knowledge held by the *intelligentsia* of Italy in the early fourteenth century,[1] that is, on the eve of the Renaissance.

On collating the many scattered references germane to our study, we discover that Dante's system of geography was to a large extent a 'harmony' of ideas advanced by such men as Brunetto Latini (his contemporary and a notorious plagiarist of Solinus), Orosius, Alfragan, Albert the Great and Isidore.[2] Thus, Dante considered the *oikoumenē* to be ocean-girt [3] and to be confined to the northern hemisphere. 'I believe,' he declares, 'that this Circle (i.e. the equator) would divide this land which is not covered by the great Ocean, yonder to the south, almost along the whole edge of the first zone (i.e. of habitable land).' [4]

Elsewhere he describes the southern hemisphere as 'mondo senza gente'.[5] The notion of an antipodal race he censures on the double ground that the Ancients, as well as Christians, would be averse to believing in a polygenetic origin for mankind. 'If Aristotle will forgive the expression, one may well call those [people] asses who can believe any such thing as this!' [6] In seeking to explain this contrast between the two hemispheres, Dante indulges in the speculation (presumably original) that when Lucifer fell from heaven, he alighted in the southern hemisphere and the

[1] Dante was born in 1265 and died in 1321.

[2] *Vide* P. Toynbee: *Dante—Studies and Researches*, pp. 38 *et seq.*, and E. Moore: *Studies in Dante* (Third Series), pp. 110 *et seq.*

[3] *Paradiso*: IX, 84. 'quel mar che la terra inghirlanda'. Cf. Brunetto's *Il Tesoro*, I, 122.

[4] *Convivio*: III, v. [5] *Inferno*: XXVI, 117. [6] *Convivio*: IV, xv.

land that was formerly there, for fear of him, covered itself with the sea.[1] When his abode became the caverns of the earth, the earth that was there fled away, and, leaving a hollow space, rushed up so as to form the solitary Mount of Purgatory.[2]

But Dante is usually much less original in his geographical thinking. For instance, he will have nothing to do with those of his contemporaries who affirm that the watery circumference of the globe is at some points[3] higher than the dry land. In his *Quaestio de aqua et terra*, a Latin lecture delivered in Verona in 1320, he shows that the round sheet of water which covers most of our sphere must have the same centre as the earth which it envelops; that the surface of the water is always at an even distance from the centre and that if the earth, which is heavier than water, rises above it in a region, the cause is not to be sought among natural laws, because the phenomenon is contrary to the law of gravitation. It is the stars, he submits, which by their combined attraction have pulled a part of the underlying earth above the water. Why God ordained that this should take place in the northern, rather than in the southern hemisphere is a mystery which it would be presumptuous to try and probe. Did not Job cry 'Canst thou by searching find out God?'[4]

As to the size of the earth, Dante adopted the estimate of 20,400 miles which was based upon Alfragan's *data* of $56\frac{2}{3}$ miles to a degree and 4,000 cubits to a mile,[5] for he tells us that Rome is 2,700 miles from the North Pole and 7,500 miles from the South Pole, which gives 10,200 miles for the semi-circumference of the globe.

On the extent of the dry land ('gran secca'), Dante declares that its longitude is such that 'at the equinox the sun is setting for those who are at one of these boundaries, when it is rising for

[1] *Inferno*: XXXIV, 121–4.

[2] This was the only dry land in the southern hemisphere, but as its inhabitants were no longer of this earth, it was still true to say that that hemisphere was 'mondo senza gente'.

[3] *Vide ante*, pp. 216–7.

[4] *passim.*

[5] Those, e.g. Sacrobosco in his *climata* and Pierre d'Ailly in his *Imago Mundi*, who regarded these cubits as being *short* cubits of fifteen inches obtained the figure of 20,400 miles, but those—the majority of Alfragan's followers—who held them to be *great* cubits of eighteen inches obtained the much more accurate value of 24,480 miles.

those who are at the other '.[1] This obviously involves a difference of 180°—a somewhat conservative estimate judged by medieval standards.[2] Concerning the subject of latitude, the same work [3] goes on to say that 'it extends from those whose zenith is a circle described from (i.e. through) the pole of the zodiac round the pole of the earth as a centre (i.e. the Arctic Circle)': in other words, about 67° in extent. The writer then concludes: 'so it is evident that the emergent earth or dry land must have the appearance of a half-moon or thereabouts'.[4]

On the descriptive side, Dante tells us little or nothing that is new. The only inhabitants of tropical Africa he appears to have heard of are the Garamantes, spoken of by all medieval writers. These 'live under the equator, having constant equality of day and night, and suffering from such excessive heat that they cannot endure any clothing'.[5] At the northern extremity of his habitable zones, in the neighbourhood of the Riphaean Mountains,[6] he places the equally venerable Scythians who 'suffer extreme inequality of days and nights and are oppressed with intolerable cold'.[7] Eastwards the River Ganges and westwards the 'isles of Gades' comprise the limits of the known world, exactly as they had done for Orosius and many another earlier writer.[8] At the very heart and centre of Dante's 'gran secca' lies Jerusalem,[9] in keeping with current belief,[10] while Rome is placed midway between Jerusalem and the western extremity of Europe.[11]

The Earthly Eden Dante describes as being situated upon a lofty mountain, beyond the reach of all atmospheric disturbances, so as to be fit for the eternal abode of man. This idea is also common in the writings of the time.[12] But in placing it in the *southern* hemisphere on the only piece of dry land found there, our author departs from the orthodox tradition of his day, which

[1] *Quaestio de aqua et terra*, par. 19.

[2] *Vide ante*, pp. 86–7.

[3] Now generally regarded as genuine.

[4] Cf. Ristoro of Arezzo: *La composizione del mondi*, VI, 11.

[5] *De Monarchia*: I, xiv.

[6] Cf. Orosius: *Historia*, I, 2, par. iv.

[7] *De Monarchia*: I, xiv.

[8] Cf. Orosius: *Historia*, I, 2, pars. vii, xiii.

[9] *Vide Purgatorio*: XV, 1–6.

[10] *Vide ante*, p. 186.

[11] *Vide Purgatorio*: II, 1–9; XXVII, 1–6.

[12] According to the popular medieval belief, atmospheric disturbances were caused by diabolic agency, i.e. by 'the prince of the power of the air'.

located it in the East.[1] All the same, he does not hesitate to locate the Tigris and Euphrates Rivers there.[2] Had he been asked to explain how these rivers could be in two different landmasses at one and the same time, no doubt he would have resorted to the hypothesis (first propounded by St. Augustine) that they flowed underground to the spot where they first came to the surface, and where they were therefore supposed to rise.[3]

When he comes nearer home, Dante exhibits no inconsiderable knowledge of regional topography and hydrography, particularly the latter, for whenever he can possibly describe a country or town allusively by reference to its river he does so. Thus he speaks of London as 'in sul Tamigi';[4] Paris, Florence, Lucca, Bologna are likewise spoken of periphrastically under the name of their rivers. Another allusion which suggests that Dante was more than usually acquainted with European geography comes where he is describing the embankment on the borders of the river Phlegethon in Hell and compares it to the dykes built by the Flemings along the sea-coast between Guizzante (i.e. Wissant near Calais) and Bruges.[5]

If we are tempted to reflect that, in all this, Dante has said no word about contemporary discovery, that he does not appear to have heard of Cathay,[6] or the travels of his fellow-countryman, Marco Polo, we need to remember that few men of his day, even those who were professedly writing on geographical subjects, tell us much of what was going on around them; further, that even if he knew of them, neither his *Divina Commedia* nor his lesser writings gave him the necessary excuse for treating of them. As it is, his works, prose and poetical alike, provide us with a very valuable commentary on the geographical lore of the later years of the Middle Ages.

[1] *Vide* Plates No. III, XV. Later on, in maps of the second half of the fourteenth century and the fifteenth century, Paradise was sometimes located in central and *south* Africa; *vide* Plates No. IX, X, XI.

[2] *Vide Purgatorio*: XXXIII, 112.

[3] *Vide ante*, pp. 173–4.

[4] *Inferno*: XII, 120.

[5] *Ibid.*: XV, 4–6.

[6] He was, however, familiar with Tartar cloth (*vide Inferno*: XVII, 14–17), but as this term was used in the Middle Ages generically of all rich stuffs of Oriental origin, it does not follow that Dante knew anything of the lands of Tartary.

SELECT BIBLIOGRAPHY

ABBREVIATIONS USED IN BIBLIOGRAPHY

Migne: *Pat. lat.* . . .	J. P. Migne: *Patrologiae cursus completus . . . series latina*, 221 vols., Paris (1844–64).
Migne: *Pat. graec.* . . .	J. P. Migne: *Patrologiae cursus completus . . . series graeca.*
Bibl. geogr. arab. . . .	*Biblioteca geographorum arabicorum.* Edited by M. J. de Goeje, Leiden, 1870–94.
Corp. script. eccl. lat. . . .	*Corpus scriptorum ecclesiasticorum latinorum*, Vienna, 1866 *et seq.*
Script. rerum. germ. in usum schol. .	*Scriptores rerum germanicarum in usum scholarum ex monumentis Germaniae historicis recusi*, 1840 *et seq.*
Geogr. graec. min. . . .	*Geographi graeci minores.* Edited by C. Müller, Paris, 1882.

I. TEXTS

Abano, Peter of. *Conciliator differentiarum*, Venice, 1496.

Abelard, Peter. *Expositio in hexaemeron*, in Migne: *Pat. lat.*, vol. CLXXVIII.

Abū'l Fidā. *Géographie d'Aboulfeda* (French text and commentary by J. T. Reinaud), Paris, 1848.

Achillini, Alessandro. *De elementis.* Venice, 1508.

Adam of Bremen. *Gesta Hammaburgensis ecclesiae pontificum*, in Migne: *Pat. lat.*, vol. CXXXXVI.

Adam, William (?), Archbishop of Soltania. *Directorium ad faciendum passagium transmarinum.* Edited by C. R. Beazley in *American Historical Review*, vols. XII, XIII, 1907.

Adelard of Bath. *Quaestiones naturales.* Translation by Gollancz under title of *Dodi ve-Nechdi*, 1920.

Ibid. *De eodem et diverso.* Edited by H. Willner in *Beiträge zur Geschichte der Philosophie des Mittelalters*, vol. IV, pt. 1, Münster, 1903.

Aeneas Silvius. *Vide* Pius II.

Agobard. *Liber contra insulsam vulgi opinionem de grandine et tonitruis*, in Migne: *Pat. lat.*, vol. CIV.

Ailly, Pierre d'. *Tractatus de imagine mundi*, Louvain ?, 1480 ?.

Albert the Great. *Opera omnia.* Edited by A. Borgnet, Paris, 1890–9.

Albert of Saxony. *Questiones subtilissime Alberti de Saxonia in libros de celo et mundo*, Venice, 1492.

Albertus Magnus. *Vide* Albert the Great.

Alfragan (Al-Farghānī). *Rudimenta astronomica.* Translation by J. Hispalensis, Paris, 1546.

Ambrose, St. *Opera.* Edited by C. Schenkl, Vienna, 1896–1919.

Anon. Ravenna Geographer. *Cosmographia.* Edited by M. Pinder and G. Parthey in *Ravennatis anonymi cosmographia.*, Berlin, 1860.

Anon. *Konungs-Skuggsjá.* Translation by L. M. Larson, New York, 1917.

Anon. *De imagine mundi* (c. 1100), in Migne : *Pat. lat.*, vol. CLXXII, where it is attributed to Honorius of Autun.

Anon. *Libro del conoscimiento de todos los reynos y tierras.* Translation by C. R. Markham (Hakluyt Society), 1912.

Aristotle. *De caelo.* Translation by J. L. Stocks, Oxford, 1922.

Ibid. Meteorologica. Translation by E. W. Webster, Oxford, 1923.

Augustine, St. *De civitate dei* and *Soliquiorum libri duo*, in Migne : *Pat. lat.*, vols. XXXII–XLVII.

Avienus, Rufus. *Descriptio orbis terrae, Ora maritima . . . (Bibliotheca classica latina, poetae latini minores*, vol. 5), Paris, 1825.

Bacon, Roger. *Opus Majus.* Edited by J. H. Bridges. Oxford, 1897–1900. Translation by R. B. Burke, Philadelphia, 1928.

Bakrī, Al-. *Book of Roads and Kingdoms.* Edited by F. Wüstenfeld, Göttingen and Paris, 1876–7.

Balkhī, Al-. *Figures of the Climates.* Edited and translated by G. Le Strange and R. A. Nicholson. Cambridge, 1921.

Bandinus (Dominicus) of Arezzo. *Fons memorabilium universi* (Balliol College Library, MS. No. 238).

Bartholomew (Anglicus) of England. *De proprietatibus rerum*, Nürnberg, 1492.

Basil, St. *Hexaemeron*, in Migne : *Pat. graec.*, vol. XXIX.

Battānī, Al-. *On the Movements of the Stars.* Latin translation in C. A. Nallino : *Al-Battānī sive Albatenii opus astronomicum*, Milan, 1899–1907.

Batūta, Ibn. *Travels in Asia and Africa.* Translation by H. A. R. Gibbs, London, 1929.

Beatus, St. *Beati in Apocalipsin libri duodecim.* Edited by H. A. Sanders. Rome, 1930.

Bede (the Venerable). *De natura rerum* and *De temporum ratione.* Basle, 1529. English translation by J. A. Giles, London, 1843.

Benjamin of Tudela. *The Itinerary of Benjamin of Tudela.* English translation by M. N. Adler, London, 1907.

Bernard Sylvester. *De mundi universitate.* Edited by C. S. Barach and J. Wrobel in *Biblioteca philosophorum mediae aetatis*, vol. I, Innsbruck, 1876.

Bernard of Verdun. *Tractatus optimus super totam astrologiam, vide* Duhem, P.

Bethencourt, Jean de. *The History of the Conquest of the Canaries.* Translation by R. H. Major (Hakluyt Society), 1871.

Bīrūnī, Al-. Geographical writings translated and edited by J. T. Reinaud: *Géographie d'Aboulfeda*, Paris, 1848.

Boethius. *De consolatione philosophiae.* Translation by G. Colville, London, 1897.

Brunetto Latini. *Vide* Latini, Brunetto.

Burchard of Mount Sion. *Descriptio terrae sanctae.* Translation by A. Stewart (Palestine Pilgrims' Text Society), vol. XXX. London, 1896.

Ca'da Mosto, Luigi di. *Le navigazioni atlantiche de Alvise da Ca'da Mosto.* Edited by R. Caddeo, Milan, 1929.

Calderia, John (of Venice). *Liber canonum astrologiae ac totius orbis descriptione* (sic) . . . Venice, S. Marco Library, XX, 95 (Valentinelli, XI, 93).

Capella, Martianus. *Satyricon* or *De nuptiis philologiae mercurii.* Edited by F. Eyssenhardt, Leipzig, 1866.

Cassiodorus. *De artibus et disciplinis liberalium artium* and *Institutiones divinarum et humanarum literarum*, Paris, 1589.

Castro, João de. *Roteiro de Lisboa a Goa*, Lisbon, 1538.

Caxton, William. *Vide* Gossouin of Metz.

Clavijo, Ruy G. *Narrative of the Embassy of Ruy Gonzalez de Clavijo to the Court of Timour.* Translation by C. R. Markham (Hakluyt Society), 1859.

Columbus, Christopher. *Select Letters of Columbus.* Translation by R. H. Major (Hakluyt Society), 1847.

Ibid. The Journal of Christopher Columbus. . . . Translation by C. R. Markham (Hakluyt Society), 1893.

Ibid. Voyages of Columbus. Translation by C. Jane (Hakluyt Society), 1929–32.

Comestor, Peter. *Historia scholastica*, in Migne: *Pat. lat.*, vol. CXCVIII.

Constantine VII (Porphyrogennetos). *De administrando imperii*, in Migne: *Pat. graec.*, vol. CXIII.

Cosmas Indicopleustes. *Christian Topography.* Translation by J. W. McCrindle (Hakluyt Society), 1897.

Dante Alighieri. *Quaestio de aqua et terra* (not certainly the work of Dante). Translation by C. L. Shadwell, Oxford, 1909.

Ibid. Divina Commedia. Translation by C. E. Norton, Boston, 1892–3.

Ibid. Convivio (*Convito*). Translation by P. H. Wicksteed, London, 1903.

Dati, Leonardo. *La sfera* (Brit. Mus., Add. MSS. 22,329 and 24,942).

Dicuil. *De mensura orbis terrae.* Edited by A. Letronne in his *Recherches géographiques et critiques sur le livre De mensura orbis terrae*, Paris, 1814.

Edrisi. *Vide* Idrīsī, Al-.

Faleiro, Francisco. *Arte del marear.* Lisbon, 1532.

FONTANA, GIOVANNI DA. *Liber de omnibus rebus naturalibus quae continentur in mundo . . .*, Venice, 1554.

Ibid. Libellus de Speculo mikesi (Vatican Library, Barberini MS., 350).

GALVÃO, ANTONIO. *Discoveries of the World . . .* Edited by C. R. D. Bethune (Hakluyt Society), 1863.

GERALD OF BARRY. *Giraldi cambrensis opera* (Rolls Series, No. 21). Edited by J. S. Brewer, London, 1861. English translation by T. Foster, London, 1863.

GERBERT. *Vide* Sylvester II.

GERVASE OF CANTERBURY. *Mappamundi* in *The Historical Works of Gervase of Canterbury* (Roll Series, No. 71). Edited by W. Stubbs, London, 1889.

GERVASE OF TILBURY. *Otia imperialia.* Edited by G. Leibnitz, Hanover, 1707–11.

GIRALDUS CAMBRENSIS. *Vide* Gerald of Barry.

GOSSOUIN OF METZ (?). *Image du monde.* Translation by William Caxton —*The Mirrour of the World*—London, 1485.

GREGORY OF NAZIANZUS. *In hexaemeron liber*, in Migne: *Pat. graec.*, vol. XLIV.

GROSSETESTE, ROBERT. *Opera.* Edited by L. Baur in *Die philos. werke des Robert Grosseteste, Bischofs von Lincoln*, Münster, 1912.

GUIDO. *Geographica.* Edited by M. Pinder and G. Parthey in *Ravennatis anonymi cosmographia et Guidonis geographica*, Berlin, 1860.

HAUKAL, IBN. *Book of Roads and Provinces.* Latin translation by M. J. de Goeje, Leiden, 1873.

HIGDEN, RANULF. *Polychronicon.* Translation by William Caxton, London, 1480.

HILDEGARD OF BINGEN. *Subtilitates diversarum naturarum creaturarum*, in Migne: *Pat. lat.*, vol. CXCVII.

HONORIUS OF AUTUN (?). *De imagine mundi*, in Migne: *Pat. lat.*, vol. CLXXII.

HOVEDEN, ROGER. *Chronica.* Edited by W. Stubbs (Rolls Series, No. 51), London, 1868–71.

HUGO OF ST. VICTOR. *De sacramentis christianae fidei*, in Migne: *Pat. lat.*, vol. CLXXVI.

IDRĪSĪ, AL-. *La géographie d'Edrisi.* Translation by P. A. Jaubert, Paris, 1836–40.

ISTAKHRĪ, AL-. *Routes of the Provinces.* Edited by M. J. de Goeje in *Bibl. geogr. arab.*, Leiden, 1870.

ISIDORE OF SEVILLE. *Etymologiae sive originum libri XX.* Edited by W. M. Lindsay, Oxford, 1911.

Itineraries. (i) *Itinerarium provinciarum Antonini Augusti.* French translation by A. Aures, Nimes, 1868.

(ii) *The Bordeaux–Jerusalem Itinerary.* Translation by A. Stewart (Palestine Pilgrims' Text Society), London, 1887.

(iii) *The Pilgrimage of Etheria.* Translation by M. L. McClure and C. L. Feltoe, London, 1920.

JACOB BEN MACHIR. *Quadrans judaicus* or *Quadrans novus.* (Bibl. Nat., Paris; Latin MSS., No. 7,437).

JACQUES DE VITRY. *The History of Jerusalem, A.D. 1180.* Translation by A. Stewart (Palestine Pilgrims' Text Society), London, 1897.

JAIME, DUKE OF BRAGANÇA (editor). *Alguns documentos do archivo nacional da Torre do Tombo acerca das navegações e conquistas Portuguezas*, Lisbon, 1892.

JEROME, ST. *Opera omnia.* Edited by Erasmus, Basle, 1565.

JOHN OF DONDIS (DONDI). *Aggregator Paduanus de medicinis simplicibus*, Venice, 1481.

JOHN OF GLOGAU. *Introductorium compendiosum in tractatum sphere materialis magistri Joannis de Sacrobusto*, Cracow, 1506.

JOHN OF HESSE. *Itinerarius Johannis de Hese, presbiteri, a Jherusalem per diversas mundi partes . . .* Cologne, 1490.

JOHN OF PLANO CARPINI. *The Journey of . . . John of Pian del Carpine.* Translation by W. W. Rockhill (Hakluyt Society), 1900.

JOHN OF WÜRZBURG. *Description of the Holy Land.* Translation by A. Stewart (Palestine Pilgrims' Text Society), London, 1890.

JORDANUS OF SEVERAC. *Marvels described by Friar Jordanus . . .* Translation by H. Yule (Hakluyt Society), 1863.

JUBAYR, IBN. *The Travels of Ibn Jubayr.* Edited by M. J. de Goeje, Leiden, 1907.

KHURDĀDHBIH, IBN. *Livre des routes et des provinces.* Translation by M. J. de Goeje in *Bibl. geogr. arab.*, Leiden, 1889.

KHWĀRIZMĪ, AL-. *The Khorazmian Tables* in H. Suter: *Die astronomischen Tafeln des Muhammed ibn Mūsā al-Khwārizmī . . .*, Copenhagen, 1914.

LACTANTIUS, FIRMIANUS. *Divinae institutiones.* Edited by S. Brandt in *Corp. script. eccl. lat.*, vol. XIX, Vienna, 1890.

LAMBERT OF ST. OMER. *Liber floridus*, in Migne: *Pat. lat.*, vol. CLXIII.

LATINI, BRUNETTO. *Il Tesoro di M. Brunetto latino . . .* Venice, 1533.

LEO AFRICANUS. *The History and Description of Africa done into English by John Pory.* Edited by R. Brown (Hakluyt Society), 1896.

LULL, RAMON. *Arbor scientiae.* Lyons, 1515.

Ibid. *Blanquerna.* Translation by E. A. Peers, London, 1923.

MACROBIUS, AMBROSIUS. *In somnium Scipionis commentarius.* Edited by J. M. N. D. Nisard, Paris, 1883.

MADRIGNANO, ARCHANGELO. *Itinerarium Portugallensium.* 1508.

MANDEVILLE, SIR JOHN. *Mandeville's Travels.* Edited by J. P. Hamelius (Early English Text Society), London, 1919–23.

MARCELLINUS, AMMIANUS. *Rerum gestarum libri XXXI.* English translation by C. D. Yonge, London, 1862.

MARIGNOLLI, JOHN. *Cronica . . .* Translation by H. Yule in *Cathay and the Way Thither*, vol. III (Hakluyt Society), 1914.

MAS'ŪDĪ, AL-. *Les prairies d'or.* French edition by C. B. de Meynard and P. de Courteille. Paris, 1861–77. English translation (unfinished) by A. Sprenger, London, 1841.
MAUR, RABAN. *De universo,* in Migne : *Pat. lat.,* vol. CXI.
MELA, POMPONIUS. *De situ orbis.* Venice, 1471. French edition by J. M. N. D. Nisard, Paris, 1883.
MERLE, WILLIAM. *Consideraciones temperiei pro 7 annis 1337–44.* English translation by G. J. Symons, London, 1891.
MUQADDASĪ, AL-. *The Best of Classification for the Knowledge of Climate.* Edited by M. J. de Geoje in *Bibl. geogr. arab.,* Leiden, 1877.
NECKAM, ALEXANDER. *De naturis rerum libri duo.* Edited by T. Wright (Rolls Series, No. 34), London, 1863.
NIGER, MARIUS (DOMINICUS). *Geographiae commentarius . . .* Basle, 1557.
NUNES, PEDRO. *Tratado em defensam da carta de marear,* Lisbon, 1537.
OCCAM, WILLIAM OF. *Vide* William of Occam.
ODORIC OF PORDENONE. *Descriptio orientalium partium . . .* Edited by H. Yule : *Cathay and the Way Thither,* vol. II (Hakluyt Society), 1913.
ORESME, NICHOLAS. *De configuratione qualitatum . . .,* Venice, 1505.
OROSIUS, PAULUS. *Historiarum adversus paganos libri VII.* Edited by C. Zangemeister, Leipzig, 1889. English translation by I. W. Raymond (Early English Text Society), 1883.
OTTO, BISHOP OF FRIESINGEN. *Ottonis Frisingensis chronicon.* Edited by A. Hofmeister in *Script. rerum germ. in usum schol.,* Hanover and Leipzig, 1912.
PACHECO, DUARTE. *Esmeraldo de situ orbis.* English translation by G. H. T. Kimble (Hakluyt Society), 1937.
PARIS, MATTHEW. *Chronica majora.* Edited by H. R. Luard (Rolls Series, No. 57), London, 1872–83.
PAUL THE DEACON. *Historia gentis Langobardorum.* English translation by W. D. Foulke, New York, 1907.
PEGOLOTTI, F. B. *Libro di divisamenti di paesi.* Translation by H. Yule : *Cathay and the Way Thither,* vol. III (Hakluyt Society), 1914.
PETAHIAH BEN JACOB. *Travels of Rabbi Petachia.* Translation by A. Benish and W. F. Ainsworth, London, 1856.
PHOCAS, JOANNES. *Brief Description of the Castles and Cities from Antioch to Jerusalem.* Translation by A. Stewart (Palestine Pilgrims' Text Society), London, 1889.
PIGAFETTA, ANTONIO. *Il primo viaggio intorno del mondo,* Milan, 1800.
PIUS II. *Opera geographica et historica.* Helmstadt, 1699.
PLANO CARPINI, JOANNES. *Vide* John of Plano Carpini.
PLINY, C. SECUNDUS. *Natural History.* Translation by J. Bostock and H. T. Riley, London, 1855–7.
POLO, MARCO. *The Book of Ser Marco Polo.* Translation by H. Yule and H. Cordier, London, 1903 (revised in 1920).

POLO, MARCO. *The Travels of Marco Polo . . .* Edited by N. M. Penzer, London, 1929.
PRISCIAN. *Periegesis.* Edited by C. Müller in *Geogr. graec. min.*, vol. II, Paris, 1882.
PTOLEMY, CLAUDIUS. *Geographia.* Rome, 1507. English translation by E. L. Stevenson, New York, 1932.
Ibid. *Almagest.* Edited by J. L. Heiberg in *Claudii Ptolemaei opera quae exstant omnia*, vol. I, Leipzig, 1898–1903.
RABAN MAUR. *Vide* Maur, Raban.
RAVENNA GEOGRAPHER. *Vide* Anon. Ravenna Geographer.
REGIOMONTANUS, JOANNES. *Ephemerides.* Augsberg ?, *c.* 1474.
RICOLD OF MONTE CROCE. *Itinerarium . . .* Edited by J. C. M. Laurent in *Peregrinatores medii aevi quatuor*, Leipzig, 1864.
RISTORO OF AREZZO. *La composizione del mondi.* Edited by Enrico Narducci, Rome, 1859.
ROBERT OF LINCOLN. *Vide* Grosseteste, Robert.
ROBERT OF YORK. MS. (untitled) in Camb. Univ. Lib. (Ii, i, 1).
ROTZ, JAN. *The Boke of Ydrography . . .* (Brit. Mus., Cotton MS. 20 E.ix), 1542.
RUBRUCK, WILLIAM OF. *Vide* William of Rubruck.
SACROBOSCO, JOANNES (JOHN OF HOLYWOOD). *Tractatus de sphaera*, Venice, 1478.
SANUTO, MARINO. *Liber secretorum fidelium crucis super terrae sanctae recuperatione et conservatione* (Brit. Mus., Add. MSS. 17,376).
SAXO GRAMMATICUS. *The Nine Books of the Danish History.* Translation by O. Elton, London, 1894.
SCHILTBERGER, JOHANN. *Reisebuch.* Edited by V. Langmantel, Tübingen, 1885. English translation by J. B. Telfer (Hakluyt Society), 1879.
SCOT ERIUGENA, JOHN. *De divisione naturae*, in Migne: *Pat. lat.*, vol. CXXII.
SCOT, MICHAEL. *Liber particularis* (Bodleian, Canon. Misc. 555).
SENECA. *Quaestiones naturales.* English translation by J. Clarke, London, 1910.
SIMOCATTA, THEOPHYLACTUS. *Theophylacti Simocattae expraefecti et observatoris coactorum historiarum libri VIII . . .* Paris, 1647.
SOLINUS, C. JULIUS. *C. Iulii Solini collectanea rerum memorabilium.* Edited by T. Mommsen, Berlin, 1895.
STRABO, MARCUS. *The Geography of Strabo.* Translation by H. L. Jones, London, 1917–32.
SYLVESTER II. *Gerberti postea Silvestri II papae opera mathematica.* Edited by N. Bubnov, Berlin, 1899.
VINCENT OF BEAUVAIS. *Speculum maius (naturale, doctrinale, historiale)*, Nürnberg, 1485.
WILLIAM THE BRETON. *Gesta Philippi regis Franciae.* Edited by H. F. Delaborde in *Œuvres de Rigord et de Guillaume le Breton*, Paris, 1885.

William of Conches. *De philosophia mundi*, in Migne: *Pat. lat.*, vol. CLXXII.
Ibid. *Dragmaticon philosophiae*, Strasburg, 1567.
William of Occam. *Opera de ecclesiastica et politica potestate.* Edited by M. Goldast, Frankfürt, 1614.
William of Rubruck. *The Journey of William of Rubruck.* Translation by W. W. Rockhill (Hakluyt Society), 1900.
William of Tyre. *Historia rerum in partibus transmarinis gestarum*, in Migne: *Pat. lat.*, vol. CCI.
Ya'qūbī, Al-. *Book of the Countries.* Edited by A. W. Th. Jynboll, Leiden, 1861.
Yākūt, Ibn. *Geographical Dictionary.* Edited by C. B. de Meynard, Paris, 1871.
Yūnus, Ibn. *Le livre de la grande table Hakémite.* French edition by Caussin, Paris.
Zacuto, Abraham. *Almanach perpetuum.* Venice, 1502.
Zeno, Nicolo and Antonio. *The Voyages of the Venetian Brothers, Nicolo and Antonio Zeno, to the Northern Seas . . .* Translation by R. H. Major (Hakluyt Society), 1873.
Zurara, Eannes (Gomes) de. *The Chronicle of the Discovery and Conquest of Guinea.* Translation by C. R. Beazley and E. Prestage (Hakluyt Society), 1896–9.

II. SECONDARY MATERIAL

Andrews, M. C. 'The Study and Classification of Mediaeval Mappaemundi', *Archaeologia*, vol. LXXV, pp. 61–76.
Arnold, Sir Thomas (Editor). *The Legacy of Islam*, Oxford, 1931.
Avezac-Macaya, M. A. P. d'. 'Notice sur les anciens voyages de Tartarie en général, et sur celui de Jean de Carpin en particulier', *Recueil de voyages et de mémoirs*, vol. IV (Société de Géographie de Paris), 1839, Paris.
Ibid. 'Coup d'œuil historique sur le projection des cartes de géographie', *Bulletin de la Société de Géographie de Paris*, series 5, vol. V, 1863.
Ibid. *Notice des découvertes faites au moyen âge dans l'océan Atlantique antérieurement aux grandes explorations portuguaises du quinzième siècle*, 1845, Paris.
Babcock, W. H. *Legendary Islands of the Atlantic: A Study in Mediaeval Geography.* American Geographical Research Series, 1922, New York.
Bailey, C. (Editor). *The Legacy of Rome*, Oxford, 1928.
Barros e Sousa, Visconde de Santarem. *Recherches sur la priorité de la découverte des pays situés sur la côté occidentale d'Afrique*, 1842.
Ibid. *Essai sur l'histoire de la cosmographie pendant le moyen âge*, 3 vols., 1849–52, Paris.

BARROS E SOUSA, VISCONDE DE SANTAREM. 'Note sur la mappemonde de Fra Mauro', *Bulletin de la Société de Géographie de Paris*, series 3, 1846, Paris.
Ibid. 'Notice sur plusieurs monuments géographiques inédits du moyen âge', *Bulletin de la Société de Géographie de Paris*, series 3, 1847, Paris.
BEAZLEY, SIR RAYMOND. *The Dawn of Modern Geography*, 3 vols., 1897–1906, London.
Ibid. *Prince Henry the Navigator* (Heroes of the Nations), 1890.
Ibid. 'New Light on some mediaeval Maps', *Geographical Journal*, vols. XIV, XV, XVI, 1899, 1900.
BENSAUDE, J. *L'astronomie nautique au Portugal a l'époque des grandes découvertes*, Bern, 1912.
BETT, H. *Johannes Scotus Erigena: A Study in Mediaeval Philosophy*, Cambridge, 1925.
BETTEN, F. S. 'Knowledge of the Sphericity of the Earth during the Earlier Middle Ages', *Catholic Historical Review*, vol. III, 1923, pp. 74–90, Washington, D.C.
BEVAN, W. L., and PHILLOTT, H. W. *Mediaeval Geography, an essay in illustration of the Hereford mappamundi*, London, 1873–4.
BOER, T. J. DE. *The History of Philosophy in Islam.* Translation by E. R. Jones, London, 1903.
BOVILL, E. W. *Caravans of the Old Sahara: an Introduction to the History of the Western Sudan*, Oxford, 1933.
BOXER, C. R. 'Portuguese Roteiros, 1500–1700', *Mariner's Mirror*, vol. XX, pp. 171–86, 1934.
BREHAUT, E. *An Encyclopaedist of the Dark Ages: Isidore of Seville*, New York, 1912.
BUCHON, J. A. C., and TASTU, J. 'Notice d'un atlas en langue Catalane manuscrit de l'an 1375', *Notices et extraits des manuscrits de la Bibliothèque du Roi et autres bibliothèques*, vol. XIV, pt. ii, Paris, 1841.
BUNBURY, E. H. *A History of Ancient Geography*, 2 vols., London, 1879.
BURON, E. *Ymago Mundi de Pierre d'Ailly: texte latin et traduction française des quatre traites cosmographiques de d'Ailly et des notes marginales de Christophe Columb*, 3 vols., Paris, 1930.
BURY, J. B. *A History of the Later Roman Empire*, 2 vols., London, 1889.
CARRA DE VAUX, B. *Les penseurs de l'Islam*, 5 vols., Paris, 1921–6.
COÈDES, G. *Textes d'auteurs grecs et latins relatifs a l'extrême-Orient depuis le IVe siècle av. J.C. jusqu'au XIVe siècle*, Paris, 1910.
COLI, E. *Il paradiso terrestre dantesco*, Florence, 1897.
COOLEY, W. D. *The Negroland of the Arabs*, London, 1841.
CORDIER, H. *Les voyages en Asie au XIVe siècle du bienheureux frère Odoric de Pordenone*, Paris, 1891.
Ibid. *Documents historiques et géographiques relatifs a l'Indo-chine*, 4 vols., Paris, 1910–14.
Ibid. *Vide* Marco Polo.

CORDIER, H., and SCHEFER, C. H. *Recueil de voyages et de documents pour servir a l'histoire de la géographie depuis le XIIe jusqu'à la fin du XVIe siècle*, Paris, 1882.

CORTESÃO, A. *Cartografia e cartografos Portugueses dos seculos XV e XVI*, 2 vols., Lisbon, 1935.

DAMPIER-WHETHAM, W. *A History of Science: and its relations with Philosophy and Religion.* Cambridge, 1929.

DARBY, H. C. 'Geography in a Mediaeval Text Book', *Scottish Geographical Magazine*, vol. XLIX, pp. 323–31, 1933.

Ibid. 'The Geographical Ideas of the Venerable Bede', *Scottish Geographical Magazine*, vol. LI, pp. 84–9, 1935.

DUHEM, P. *Le système du monde: histoire des doctrines cosmologiques de Platon à Copernic*, 5 vols., Paris, 1913–17.

FERRAND, GABRIEL. *Relations de voyages et textes géographiques arabes, persans et turks relatifs à l'extrême-Orient du VIIe–XVIIe siècles*, 2 vols., Paris, 1913–14.

Ibid. *Introduction à l'astronomie nautique arabe*, Paris, 1928.

FORDHAM, SIR H. G. *Maps: Their History, Characteristics and Uses*, Cambridge, 1927.

GATHORNE-HARDY, G. M. *The Norse Discoverers of America: The Wineland Sagas translated and discussed*, Oxford, 1921.

GEIKIE, SIR A. *The Founders of Geology*, London, 1905.

GREGORY, J. W. 'The Evolution of the Map of the World', *Scottish Geographical Magazine*, vol. XXXIII, pp. 49–65, 1917.

GUNTHER, R. T. 'The Astrolabe: Its Uses and Derivations', *Scottish Geographical Magazine*, vol. XLIII, pp. 135–47, 1927.

HALLBERG, I. *L'extrême-Orient dans la littérature et la cartographie de l'Occident des XIIe, XIVe, et XVe siècles*, Göteborg, 1916.

HARRISSE, H. *The Discovery of North America: a Critical, Documentary and Historic Investigation*, London and Paris, 1892.

HASKINS, C. H. *The Renaissance of the Twelfth Century.* Cambridge, Mass., 1927.

Ibid. *Studies in the History of Mediaeval Science*, Cambridge, Mass., 1927.

Ibid. *Studies in Mediaeval Culture*, Oxford, 1929.

HEYD, W. VON. *Histoire du commerce du Levant au moyen âge*, 2 vols., Leipzig, 1885–6. (A new edition appeared in 1923, edited by F. Raynaud.)

HITTI, P. K. *History of the Arabs*, London, 1937.

HUDSON, G. F. *Europe and China*, London, 1931.

HUMBOLDT, ALEXANDER VON. *Examen critique de l'histoire de la géographie du nouveau continent*, 5 vols., Paris, 1836–9.

HUSSEY, J. M. *Church and Learning in the Byzantine Empire*, 867–1185. Oxford, 1937.

KIMBLE, G. H. T. 'The *Ne Plus Ultra* of the West African Coast in the Middle Ages', *Mariner's Mirror*, vol. XX, pp. 280–92, 1934.

KIMBLE, G. H. T. *Memoir on the Catalan World Map of the R. Biblioteca Estense at Modena* (Royal Geographical Society), 1934.
Ibid. 'The Laurentian World Map . . .', *ImagoMundi*, vol. I, pp. 29–33, 1935.
Ibid. 'Portuguese Policy and its Influence on Fifteenth Century Cartography', *American Geographical Review*, vol. XXIII, pp. 653–9, 1933.
Ibid. 'Some Notes on Mediaeval Cartography, with special reference to Martin Behaim's Globe', *Scottish Geographical Magazine*, vol. XLIX, pp. 91–8, 1933.
KOMROFF, M. *Contemporaries of Marco Polo. . . .* London, 1928.
KRETSCHMER, K. 'Die physische erdkunde im Christlichen mittelalter' in *Geographische abhandlungen herausgegeben von Albrecht Penck*, vol. IV, pt. i, Vienna and Olmütz, 1889.
LANGLOIS, C. V. *La connaissance de la nature et du monde au moyen âge d'après quelques écrits français a l'usage des laiics*, Paris, 1911.
LELEWEL, J. *La géographie du moyen âge*, 5 vols., Brussels, 1852–7.
LETRONNE, A. 'Des opinions cosmographiques des pères de l'église, rapprochées des doctrines philosophiques de la Grèce', *Revue des deux mondes*, series 2, vol. I, Paris, 1834.
LITTLE, A. G. (Editor). *Roger Bacon Essays* . . ., Oxford, 1914.
MAJOR, R. H. *The Life of Prince Henry of Portugal, Surnamed the Navigator*, London, 1868.
MANDONNET, P. 'Les idées cosmographiques d'Albert le Grand et de St. Thomas d'Aquin et la découverte de l'Amérique', *Revue thomiste*, vol. I, St. Maximin, 1893.
MAS-LATRIE, L. DE. *Relations et commerce de l'Afrique septentrionale ou Magreb avec les nations chrétiennes du moyen âge*, Paris, 1886.
MILLER, K. *Mappaemundi, die ältesten weltkarten*, 6 vols., Stuttgart, 1895–8.
Ibid. *Mappae Arabicae: Arabische welt- und lander karten des 9–13. jahrhunderts* . . . 4 vols., Stuttgart, 1926–7.
MINORSKY, V. (Editor). *Hudūd al'Ālam* (The Regions of the World), London, 1937.
MITCHELL, J. B. 'Early Maps of Great Britain', *Geographical Journal*, vol. LXXXI, pp. 27–34, 1933.
MOORE, E. *Studies in Dante*, 4 vols., Oxford, 1896–1917.
NANSEN, F. *In Northern Mists: Arctic Explorations in Early Times*, 2 vols., New York, 1911.
NEWTON, A. P. (Editor). *Travel and Travellers in the Middle Ages*, London, 1926.
Ibid. *The Great Age of Discovery*, London, 1932.
NORDENSKIOLD, A. E. *Facsimile Atlas to the Early History of Cartography* (translation by J. A. Ekelöf and C. R. Markham), Stockholm, 1889.
Ibid. *Periplus: An Essay on the Early History of Charts and Sailing Directions* (translation by F. A. Bather), Stockholm, 1897.

NORDENSKIOLD, A. E. 'The Influence of the Travels of Marco Polo on Jacobo Gastaldi's Maps of Asia', *Geographical Journal*, vol. XIII, pp. 396–406, 1899.

NUNN, G. E. *The Geographical Conceptions of Columbus*. American Geographical Research Series, New York, 1924.

PAETOW, L. J. *A Guide to the Study of Mediaeval History*, London, 1931.

POOLE, R. L. *Illustrations of the History of Mediaeval Thought and Learning*, London, 1920.

PRESTAGE, E. *Portuguese Pioneers*, London, 1933.

RAINAUD, A. *Le continent austral : hypothèses et découvertes*, Paris, 1893.

RAND, E. K. *Founders of the Middle Ages*, Cambridge, Mass., 1929.

RAVENSTEIN, E. G. *Martin Behaim : His Life and His Globe*, London, 1908.

REINAUD, J. T. *Relation des voyages faits par les Arabs et les Persans dans l'Inde et à la Chine dans le IXe siècle de l'ère chrétienne*, Paris, 1845.

Ibid. *Vide* Abu'l Fida.

RENAN, E. *Averroès et l'Averroïsme*, Paris, 1852.

RONCIÈRE, CHARLES DE LA. *La découverte de l'Afrique au moyen âge*, 3 vols., Cairo, 1924–6.

SANDYS, SIR J. E. *A History of Classical Scholarship : From the Sixth Century B.C. to the end of the Middle Ages*, 3 vols., Cambridge, 1921.

SANTAREM. *Vide* Barros e Sousa.

SARTON, G. *Introduction to the History of Science*, 2 vols. (to date), Washington, D.C., 1927–31.

SCHOY, C. 'The Geography of the Moslems of the Middle Ages', *American Geographical Review*, vol. XIV, pp. 257–69, 1924.

SCHUTTE, G. 'Ptolemy's Atlas', *Scottish Geographical Magazine* (5 articles), vols. XXX, XXI, 1924–5.

SIMAR, T. *La géographie de l'Afrique centrale dans l'antiquité et au moyen âge*, Brussels, 1912.

SILVA, L. P. DA. *Historia da colonisação Portuguesa do Brasil*, Porto, 1921.

SINGER, C. (Editor). *Studies in the History and Method of Science*, Oxford, 1921.

STEVENS, H. *Historical and Geographical Notes on the Earliest Discoveries in America*, 1453–1530, New Haven, 1869.

STEVENSON, E. L. *Vide* Ptolemy, Claudius.

Ibid. *The Genoese World Map*, 1457 : *Facsimile and Text*, New York, 1912.

Ibid. *Terrestrial and Celestial Globes : Their History and Construction*. New York, 1921.

Ibid. *Portolan Charts : Their Origin and Characteristics*, New York, 1911.

TAYLOR, E. G. R. *Tudor Geography*, 1485–1583, London, 1930.

Ibid. 'Pactolus—River of Gold', *Scottish Geographical Magazine*, vol. XLIV, pp. 129–44, 1928.

Ibid. 'Imago Mundi', *Scottish Geographical Magazine*, vol. XLVII, pp. 78–80, 1931.

Ibid. 'Compendium Cosmographiae : A Text Book of Columbus', *Scottish Geographical Magazine*, vol. XLVII, pp. 214–19, 1931.

TAYLOR, E. G. R. 'Some Notes on Early Ideas of the Form and Size of the Earth', *Geographical Journal*, vol. LXXXV, pp. 65–8, 1935.
TAYLOR, H. O. *The Mediaeval Mind*, 2 vols., London, 1914.
THORNDIKE, LYNN. *A History of Magic and Experimental Science . . .*, 4 vols., New York, 1923–34.
Ibid. *Science and Thought in the Fifteenth Century*, Oxford, 1929.
TOYNBEE, PAGET. *Dante—Studies and Researches*, London, 1902.
TUDEER, L. O. T. 'On the Origin of the Maps attached to Ptolemy's Geography', *Journal of Hellenic Studies*, vol. XXXVII, pt. i, 1917.
VAUDREY HEATHCOTE, N. H. DE. 'Early Nautical Charts', *Annals of Science*, vol. I, pp. 13–28, 1936.
VINCENT, W. *The Commerce and Navigation of the Ancients in the Indian Ocean*, 2 vols., London, 1807.
VIVIEN DE ST. MARTIN, L. *Histoire de la géographie et des découvertes géographiques depuis les temps les plus reculés jusqu'à nos jours*, Paris, 1873–4.
WALCKENAER, C. A. *Recherches géographiques sur l'intérieur de l'Afrique septentrionale*, Paris, 1821.
WAUWERMANS, H. E. *Essai de l'histoire de l'école cartographique Anversoise du XVIe siècle*, 2 vols., Brussels, 1895.
WEST, A. F. *Alcuin and the Rise of the Christian Schools*, London, 1893.
WHEWELL, W. *History of the Inductive Sciences*, 3 vols., London, 1857.
WHITE, A. D. *The Warfare of Science with Theology*, 2 vols., London, 1895.
WHITTAKER, T. *Macrobius, or Philosophy, Science and Letters in the year 400*, Cambridge, 1923.
WIESER, F. VON. *Die weltkarte des Albertin de Virga aus dem anfange des XV. jahrhunderts*, Innsbruck, 1912.
WINTER, H. 'Who Invented the Compass?', *Mariner's Mirror*, vol. XXIII, pp. 95–102, 1937.
Ibid. 'The Pseudo-Labrador and the Oblique Meridian', *Imago Mundi*, vol. II, pp. 61–73, 1937.
WRIGHT, J. K. *Geographical Lore of the Time of the Crusades*, American Geographical Society Research Series, New York, 1925.
Ibid. 'Notes on the Knowledge of Latitudes and Longitudes in the Middle Ages', *Isis*, vol. V, pp. 75–98, 1923.
WRIGHT, T. (Editor). *Popular Treatises on Science written during the Middle Ages in Anglo-Saxon, Anglo-Norman and English*, London, 1841.
YOUSSOUF KAMAL, PRINCE. *Quelques éclaircissements épars sur mes Mon. Cart. Africae et Aegypti*, Leiden, 1935.
Ibid. *Hallucinations scientifiques (les Portulans)*, Leiden, 1937.
ZARNCKE, F. 'Der Priester Johannes' in *Abhandlungen der phil. hist. classe der K. sächsischen Gesell. d. wissenschaften*, vols. VII, VIII, 1879, 1883.
ZURLA, P. *Il mappamondo di Fra Mauro Camaldolese*, Venice, 1806.

INDEX

For Product Safety Concerns and Information please contact our EU representative GPSR@taylorandfrancis.com Taylor & Francis Verlag GmbH, Kaufingerstraße 24, 80331 München, Germany